Greenhouse Gases: A Global Overview

Greenhouse Gases:
A Global Overview

Edited by Joshua Garrett

SYRAWOOD
PUBLISHING HOUSE

New York

Published by Syrawood Publishing House,
750 Third Avenue, 9th Floor,
New York, NY 10017, USA
www.syrawoodpublishinghouse.com

Greenhouse Gases: A Global Overview
Edited by Joshua Garrett

International Standard Book Number: 978-1-68286-778-5 (Hardback)

Cataloging-in-Publication Data

Greenhouse gases : a global overview / edited by Joshua Garrett.
    p. cm.
Includes bibliographical references and index.
ISBN 978-1-68286-778-5
1. Greenhouse gases. 2. Greenhouse gases--Environmental aspects.
3. Greenhouse gas mitigation. 4. Global warming. I. Garrett, Joshua.
TD885.5.G73 G74 2019
363.738 74--dc23

TABLE OF CONTENTS

Preface.. VII

Chapter 1   A Comparative Study of Human Health Impacts due to Heavy
Metal Emissions from a Conventional Lignite Coal-Fired Electricity
Generation Station, with Post-Combustion, and Oxy-Fuel Combustion
Capture Technologies.. 1
Jarotwan Koiwanit, Anastassia Manuilova, Christine Chan,
Malcolm Wilson and Paitoon Tontiwachwuthikul

Chapter 2   Livestock as Sources of Greenhouse Gases and its Significance
to Climate Change...24
Veerasamy Sejian, Raghavendra Bhatta, Pradeep Kumar Malik,
Bagath Madiajagan, Yaqoub Ali Saif Al-Hosni, Megan Sullivan and
John B. Gaughan

Chapter 3   Effect of Dopants on the Properties of Zirconia-Supported
Iron Catalysts for Ethylbenzene Dehydrogenation with
Carbon Dioxide..41
Maria do Carmo Rangel, Sirlene B. Lima,
Sarah Maria Santana Borges and Ivoneide Santana Sobral

Chapter 4   Review of Recent Developments in CO_2 Capture using Solid
Materials: Metal Organic Frameworks (MOFs) 59
Mohanned Mohamedali, Devjyoti Nath, Hussameldin Ibrahim and
Amr Henni

Chapter 5   Mitigating Greenhouse Gas Emissions from Winter Production
of Agricultural Greenhouses...99
Lilong Chai, Chengwei Ma, Baoju Wang, Mingchi Liu and
Zhanhui Wu

Chapter 6   Greenhouse Gas Emissions – Carbon Capture, Storage and
Utilisation...117
Bernardo Llamas, Benito Navarrete, Fernando Vega,
Elías Rodriguez, Luis F. Mazadiego, Ángel Cámara and
Pedro Otero

Chapter 7   Native Forest and Climate Change — The Role of the Subtropical
Forest, Potentials, and Threats...150
Silvina M. Manrique and Judith Franco

Chapter 8   **Energy for Sustainable Development: The Energy–Poverty–Climate Nexus**...176
Melanie L. Sattler

Permissions

List of Contributors

Index

PREFACE

This book aims to highlight the current researches and provides a platform to further the scope of innovations in this area. This book is a product of the combined efforts of many researchers and scientists, after going through thorough studies and analysis from different parts of the world. The objective of this book is to provide the readers with the latest information of the field.

Greenhouse gases are the gases which absorb and emit radiant energy in the thermal infrared zone. This mechanism is the primary cause of greenhouse effect, which causes critical environmental issues and threats to ecosystems, biodiversity and human life present on Earth. Some primary greenhouse gases are methane, ozone, carbon dioxide, nitrous oxide and water vapor. The contribution of each of these gases to the greenhouse effect depends on their atmospheric lifetime, radiative forcing and global warming potential. The main causes of greenhouse gas emission are deforestation, burning of fossil fuels like oil, coal and natural gas, use of chlorofluorocarbons for refrigeration, livestock enteric fermentation, etc. This book presents researches and studies performed by experts across the globe on greenhouse gases and their effects on the environment. Different approaches, evaluations, methodologies and advanced studies on greenhouse gases have been included herein. Coherent flow of topics, student-friendly language and extensive use of examples make this book an invaluable source of knowledge.

I would like to express my sincere thanks to the authors for their dedicated efforts in the completion of this book. I acknowledge the efforts of the publisher for providing constant support. Lastly, I would like to thank my family for their support in all academic endeavors.

Editor

A Comparative Study of Human Health Impacts Due to Heavy Metal Emissions from a Conventional Lignite Coal-Fired Electricity Generation Station, with Post -Combustion, and Oxy-Fuel Combustion Capture Technologies

Jarotwan Koiwanit, Anastassia Manuilova,
Christine Chan, Malcolm Wilson and
Paitoon Tontiwachwuthikul

Additional information is available at the end of the chapter

Abstract

Carbon dioxide capture has become an important component for ensuring reduction of greenhouse gases in the atmosphere. Even though emission reduction technologies such as electrostatic precipitators (ESP) and flue gas desulfurization (FGD) are in place at most electricity-generating stations today, the large point source emitters of carbon dioxide (CO_2) and other emissions, such as heavy metals, to the atmosphere are still fossil fuel electricity-generating stations. When CO_2 capture is employed, these emissions can be further reduced. However, despite its important ability to reduce atmospheric emissions, the CO_2 capture technology in fact still releases some emissions through its stacks into the air. Since the safety and stability of the CO_2 capture technology are fundamental considerations for widespread social acceptance, the potential liability associated with the capture technology is cited as an important barrier to successful CO_2 capture implementation. Liability of the technology is further clouded by a failure to clearly define what is at risk, especially regarding human health and safety. This research study will focus on investigating the risks associated with human health and safety resulting from the different versions of the technology including: (i) no capture system, (ii) post-combustion, and (iii) oxy-fuel combustion CO_2 capture technology at the Boundary Dam Power Station (BDPS) in Estevan, Saskatchewan, Canada. The research objective of this study was to evaluate the risk to human health associated with the BDPS in Estevan, Saskatchewan, Canada, using the American Meteorological Society's Environmental Protection Agency Regulatory Model (AERMOD) and cancer and non-cancer risk equations. This research presents the air dispersion modeling of the conventional lignite-fired electricity generation station at the BDPS, the inclusion of post-combustion CO_2 capture technology, and the oxy-fuel carbon dioxide capture process. The heavy metals were measured near the power plant located in Estevan, Saskatchewan. This study shows that the emissions from the three stacks posed cancer risks of less than one chance in a million (1×10^{-6}). There were only two emissions from the "no capture" scenario that caused inhalation cancer risks of more than 1×10^{-6}. In terms of non-cancer risks, the pollutant's concentration from the three stacks was unlikely to cause any non-cancer health effects.

Keywords: carbon dioxide capture, AERMOD, air dispersion, risk, human health

---

1. Introduction

According to [1], in recent decades, climate change has had the strongest and most comprehensive impact to natural systems [2, 3]. Recent changes in climate affect heat waves, floods, wildfires, ecosystems and human systems. Emissions of CO_2 are known to contribute to the climate change as well. CO_2, a major greenhouse gas (GHG) which results in climate change, is mostly generated from electrical generation that uses fossil fuels (e.g., oil, coal, and natural gas, which are regarded as the world's primary source of energy). To cope with this problem, the use of an effective CO_2 capture technology has become an important approach in ensuring the reduction CO_2 emissions. However, since additional energy is required in carbon capture systems operation, the consumption of primary materials and fuel is increased when compared to the amount used in fossil-fuel-based energy production systems without the carbon capture technology. Consequently, it is necessary to evaluate both the energy utilization of the technology and the risks of the gaseous emissions to human health. This study focuses on the latter consideration.

The objective of this study was to analyze and compare the risks to human health posed by a lignite coal-fired electricity generation station that has the following: (i) no capture system, (ii) post-combustion, and (iii) oxy-fuel combustion CO_2 capture technology at the Boundary Dam Power Station (BDPS) in Estevan, Saskatchewan, Canada. The total area in Estevan is 795.32 square kilometers with a population density of 16.3 persons per square kilometer [4]. For the post-combustion system presented in this paper, the CO_2 is absorbed by a monoethanolamine (MEA) solvent and is purified and compressed for transportation and storage. The fuel in an oxy-fuel technology is combusted in pure oxygen (O_2) (>95% volume), which results in a concentration of CO_2 that is ready for transportation and storage. However, despite its advantages in cutting greenhouse gas (GHG) emissions, post-combustion and the oxy-fuel capture processes also emit some gases through their stacks.

A comparison of the risks to human health posed by a lignite coal-fired electricity generation station that has the following: (i) no capture system, (ii) post-combustion, and (iii) oxy-fuel combustion CO_2 capture technology at the Boundary Dam Power Station (BDPS) in Estevan, Saskatchewan, Canada, will reveal whether there are health-related risks associated with the different types of carbon capture technology. Understanding the associated risks of the technology can support formulation of the standards and regulatory frameworks required for large-scale application of the carbon capture technology [5]. In this study, the health-related risks of the three technologies are analyzed so as to shed light on the relationships between quantitative emission releases and the probability of occurrences of health effects.

This paper is organized as follows: Section 2 presents some background to the study and provides a discussion on health effects of selected power plant pollutants, Section 3 presents methods of LCA, Section 4 provides several methods for air dispersion modeling and risk assessment of post- and oxy-fuel combustion CO_2 capture processes, Section 5 discusses the results from the analysis, Section 6 gives the discussion, and Section 7 presents conclusion and discusses some direction for future work.

2. Background health effects from typical power plants

2.1. Background and related work

To assess the emissions from the stack and the environmental impacts of the carbon capture technology, three case scenarios of a typical power plant were evaluated. The three scenarios include a power plant with the following: (i) no carbon capture system, (ii) the post-combustion carbon capture system, and (iii) the oxy-fuel combustion carbon capture system. The life cycle inventory (LCI) results generated from a life cycle assessment (LCA) study were used for calculating the pollution concentrations in each grid block within the plume area [6–8]. Air dispersion modeling has been used to evaluate the concentration in each grid block. After that, the concentrations are evaluated for the possible impacts on human health. The emissions released from the tall stacks of the electricity generation plants were not deposited near the source, but further away [9, 10]. $PM_{2.5}$ is ingested into the body via the respiratory system. Hg^0 has the longest atmospheric life span of the various species of mercury and can be transported easily over long distances due to its insolubility in and low reactivity to water. Hg^0 is the common mercury species in lignite [11]. Hg^p and Hg^{2+}, with their high reactivity and solubility in water, can be controlled by some emission control units such as electrostatic precipitators (ESP) and wet and dry flue gas desulfurization (FGD) [10, 12]. In addition, while rainfall parameters (e.g., wind, temperature, inversions, rainfall's duration, frequency, and intensity) and precipitation near the stacks affect the deposition of wet mercury (Hg), various meteorological factors such as wind speed affect the deposition of dry Hg [12, 13]. According to [9] and [14], even though most power plants were unlikely to cause any significant non-cancer risks to human health, arsenic (As), chromium (Cr), and lead (Pb) were the primary contributors to these risks. For cancer risks, the results showed that the pollutants would not cause any carcinogenic health effects to the population [9, 14]. The studies on air dispersion and risks from coal-fired power plants are summarized in **Table 1**.

Study	Country	Air dispersion and risk methods	Technology/power plant	Results
	Taiwan	ISCST	- 550 MW coal -fired power plant with ESP, FGD, and SCR - 10 stacks	- The average gaseous Hg (Hg^0 and Hg^{2+}) was 2.59–4.12 ng/m^3 - The average particulate Hg (Hg^p) was 105–182 pg/m^3 - The majority of the Hg from the stacks was in gaseous form, so the particulate form was very low - The maximum concentration of total Hg was from downwind site D (10 km from the plant) - The lowest concentration of total Hg was from upwind site A (11 km from the plant)
Lee and Keener Table [12]	USA	AERMOD and ISCST3	- 2 coal-fired power plants - 4 stacks for each plant	- The average annual atmospheric mercury concentration was 0.014–0.085 ng/m^3 depending on each power plant and air dispersion modeling - The average annual dry Hg deposition was 3.62–6.25 µg/m^2 depending on each power plant and air dispersion modeling - The average annual dry Hg deposition was 0.35–13.73 µg/m^2 depending on each power plant and air dispersion modeling - Wet Hg deposition is influenced by rainfall parameters and precipitation near the stacks - Dry Hg deposition depends on meteorological factors - There were similar trends of Hg deposition between these two power plants
Mokhtar et al.	Malaysia	AERMOD and quality health	- 700 MW coal -fired power plant	- The predicted atmospheric As, Cd, Cr, and

Study	Country	Air dispersion and risk methods	Technology/power plant	Results
Table [14]		risk assessment (QHRA)	with ESP and FGD - 3 power plants	Pb concentrations were 1.84×10^{-4}, 2.3×10^{-5}, 5.38×10^{-4}, 1.73×10^{-4} $\mu g/m^3$
				- Hazard quotient (HQ) values of all pollutants concentration were less than one. This showed that the pollutants concentration were unlikely to cause any non-cancer risks to human health
				- For cancer risks, the results showed that the pollutants would not cause any carcinogenic health effects to the population which are at 1 km away from the power plants
French et al. Table [9]	U.S.	Screening assessment	426 coal-fired and 137 oil-fired power plants	- Cancer risks: 424 of the 426 coal-fired plants did not pose any risks. As and Cr were the primary contributors to these risks
				- Non-cancer risks: None of the emissions posed these risks
				- Hg emitted during coal-fired power generation is a potential concern since it is a persistent emission which contributes to the Hg levels especially in freshwater fish. Moreover, the emission mostly does not become deposited near the source but further away

Table 1. Summary of air dispersion studies on coal-fired power plants.

2.2. Health effects of typical power plant pollutants

Emissions from a typical coal-fired electricity-generating station without carbon capture technology include secondary aerosols such as heavy metals, nitrogen oxides (NOx), sulfur dioxide (SO_2), and non-methane volatile organic compounds (NMVOC), which pose risk to human health [15]. The emissions constitute air pollution and can be hazardous to human health [3]. Health effects of selected power plant pollutants are summarized and shown in **Table 2**.

Substances	Human toxicity		Limit value		Typical exposure within the plume	Comments
	Acute (short-term effects)	Chronic (long-term effects)	TWA (the 8-hour time-weighted average (TWA) limit	STEL (short term)/C ceiling	Emission factors (kg/mg coal)	
Sulfur dioxide (SO$_2$)	Lung irritant, triggers asthma, low birthweight in infants	Reduces lung function, associated with premature death	5000 ppm	15,000 ppm	2300	Contributes to acid rain and poor visibility
Nitrogen oxides (NOx)	Changes lung function, increases respiratory illness in children	Increases sensitivity to respiratory illnesses and causes permanent damage of lung	2 ppm	5 ppm	0.054	Forms ozone smog and acid rain. Ozone is associated with asthma, reduced lung function, adverse birth outcomes, and allergen sensitization
Nitrogen dioxide (NO$_2$)	Affect health exposure mortality	Decreased lung function in children, perhaps adults	N/A	N/A	4.25	–
Carbon monoxide (CO)	Increase frequency and severity of angina, headaches, exacerbation of cardio pulmonary dysfunction	Decrease work capacity in healthy adults, decrease alertness, flulike symptom in healthy adults, asphyxiation	N/A	1 ppm	N/A	–
Particulate matter (PM)	Asthma attacks, heart rate variability, heart attacks	Cardiovascular disease, lung inflam mation, premature death, decreased lung function	25 ppm	100 ppm	N/A	Fine-particle pollution from power plants is estimated to cut short the lives of 30,000 Americans each year
Hydrogen chloride (HCl)	Inhalation causes coughing, hoarse ness, chest pain, and inflammation of respiratory tract	Chronic occupational exposure is associated with gastritis, chronic lung inflammation, skin inflammation	N/A	N/A	0.308	–

Substances	Human toxicity		Limit value		Typical exposure within the plume	Comments
Hydrogen fluoride (HF)	Inhalation causes severe respiratory damage, severe irritation, and pulmonary edema	–	N/A	2 ppm	0.6	Very high exposures through drinking water or air can cause skeletal fluorosis
Arsenic (As)	Ingestion and inhalation affect the gastrointestinal system and central nervous system	Known human carcinogen with high potency. Inhalation causes lung cancer; ingestion causes lung, skin, bladder, and liver cancer. The kidney is affected following chronic inhalation and oral exposure	N/A	2 ppm	0.075	–
Cadmium (Cd)	Bronchial and pulmonary irritation, long-lasting impairment of lung function	Human carcinogen of medium potency, kidney injury, chronic inhalation, and oral exposure	0.01 mg/m³	N/A	0.000205	Other effects noted from chronic inhalation exposure are bronchiolitis and emphysema
Lead (Pb)	Abdominal (stomach) pain, seizures	Kidney injury, decrements in renal function, anemia, paralysis, nervous system issues, and loss of cognitive ability	0.01 mg/m³	N/A	0.0000255	–
Antimony (Sb)	Gastrointestinal symptoms (vomiting, diarrhea, abdominal pain, and ulcers)	Hemolysis with abdominal and back pain	0.05 mg/m³	N/A	0.00021	Acute inhalation is related to irritation of the respiratory tract and impaired pulmonary function
Barium (Ba)	Vomiting, perioral paresthesias, diarrhea, paralysis, hypertension, and cardiac dysrhythmias	Baritosis (coughing, wheezing, nasal irritation), kidney damage	0.5 mg/m³	N/A	0.000009	The health effects depend on the dose, water solubility, and route of exposure

Substances	Human toxicity		Limit value		Typical exposure within the plume	Comments
Chromium (Cr)	High exposure to chromium VI may result in damage to the kidneys, gastrointestinal bleeding, and internal bleeding	Known human carcinogen of high potency	0.5 mg/m^3	N/A	N/A	Chronic effects from industrial exposures are inflammation of the respiratory tract, effects on the kidneys, liver, and gastrointestinal tract
Beryllium (Be)	Erythema and edema of the lung mucosa. This will produce pneumonitis	Chronic beryllium disease or berylliosis	0.5 mg/m^3	N/A	0.0000395	The effects of beryllium vary depending on the concentration of the substance in the air and the duration of the air exposure
Copper (Cu)	Nausea, vomiting, abdominal pain, anemia	Symptoms of liver toxicity such as Wilson's disease, jaundice, and swelling	0.002 mg/m^3	0.01 mg/m^3	0.0000105	–
Cobalt (Co)	Allergic contact dermatitis	Asthma, carcinogenicity	1 mg/m^3	N/A	N/A	Two routes that cobalt can be absorbed: (1) oral and (2) pulmonary routes
Molybdenum (Mo)	–	A gout-like illness, higher serum uric acid levels, carcinogenicity	0.02 mg/m^3	N/A	0.00005	–
Manganese (Mn)	–	Parkinson's disease, clumsiness, tremors, speech disturbances, psychological disturbances, cough, bronchitis, lung disease	0.5 mg/m^3	N/A	N/A	No reports of human effects following acute effects to manganese are available
Selenium (Se)	Producing coughing, nosebleeds, dyspnea, bronchial spasms, bronchitis, and chemical pneumonia	Alkali disease (hair loss, erosion of the joints of the bones, anemia, etc.), cardiovascular disease	0.2 mg/m^3	N/A	0.000245	–

Substances	Human toxicity		Limit value		Typical exposure within the plume	Comments
	(lung irritation caused by toxins, gases, etc.)					
Nickel (Ni)	Skin rash, eczema	Asthma attacks, chronic bronchitis, reduced lung function, lung, and nasal sinus cancer (>10 mg nickel/m³)	0.1 mg/m³	N/A	0.00065	People can be exposed to nickel by breathing air and drinking water
Vanadium (V)	Cough, sputum, difficulty in breathing, ear, nose, and throat irritation, headache, palpitation	Cardiovascular disease	0.05 mg/m³	N/A	0.00014	–
Mercury (Hg)	Inhalation exposure to elemental mercury results in central nervous system effects and effects on gastrointestinal tract and respiratory system	Methyl mercury ingestion causes developmental effects. Infants born to women who ingested methylmercury may perform poorly on neurobehavorial tests	0.2 mg/m³	N/A	N/A	The major effect from chronic exposure to inorganic mercury is kidney damage
Volatile organic compounds (VOCs)	Irritation, neurotoxic effects, hepatotoxic effects, headache, nausea, irritation of eyes, respiratory system, drowsiness, fatigue	Asthmatic symptom, cancer	0.025 mg/m³	N/A	0.0000415	–

Table 2. Health effects of typical coal-fired power plant pollutants (modified from Refs. Table [16–21]).

3. Methods of life cycle assessment (LCA)

LCA is a methodology that studies the whole life cycle of a product, often called the cradle-to-grave approach, in which complex systems are broken down into elementary flows. The life cycle assessment consists of four main stages: goal and scope definition, LCI analysis, life cycle impact assessment (LCIA), and interpretation. The phase of defining the goal and scope of an LCA study is important for it is at this stage that the requirements are set. The requirements determine the methodology, which can directly affect the results. The second phase of the LCA involves construction of a flow model and an inventory analysis so as to provide inventory data for supporting the goal and scope defining in the study. The LCI model is generally shown as a flowchart; and LCI modeling consists of the construction of the flowchart, data collection, and the calculation procedure [22]. The third phase of LCIA aims to specify the environmental consequences in the inventory analysis process. This phase is normally applied to translate the environmental load, inputs, and outputs, based on the inventory results, into environmental impacts such as acidification, global warming potential, and ozone depletion. The last stage of an LCA is the interpretation of outcomes. At this stage, the main objectives include reaching conclusions and preparing recommendations for action. The conclusion should also be consistent with the goal and scope of the study.

The study focuses on using the emission outputs from the LCI step for calculating the emission concentration using air dispersion modeling. Then, the results are used to generate the cancer and non-cancer risks. All unit processes in each scenario of the carbon capture technology are modeled using engineering equations incorporated in a Microsoft® Excel spreadsheet.

4. Methods of air dispersion modeling and risk assessment of post- and oxy-fuel combustion CO_2 capture technologies

4.1. The selected technological boundaries

To assess health-related risks due to heavy metals, three scenarios are compared, which include (i) the conventional lignite-fired electricity generation station without CO_2 capture, (ii) the amine post-combustion CO_2 capture system, and (iii) the oxy-fuel combustion CO_2 capture

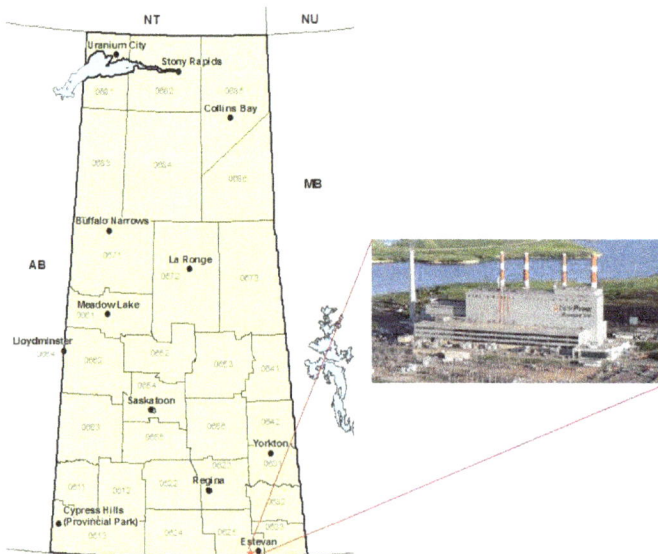

Figure 1. Boundary Dam Power Station (BDPS) in Estevan, Saskatchewan, Canada.

system. The lignite-fired electricity generation station at the BDPS in Estevan, Saskatchewan, Canada, was used in this study; the BDPS is shown in **Figure 1** [23, 24].

The three technologies are compared. These technologies include the following: (i) the conventional lignite-fired electricity generation station without CO_2 capture, (ii) the lignite coal-fired electricity-generating unit with an amine-based post-combustion capture system, and (iii) the oxy-fuel combustion CO_2 capture system. Each technology is described as follows. The conventional lignite-fired electricity generation station consists of (i) unit 3 at the BDPS, which generates 150 MW and is a tangentially fired subcritical boiler, and (ii) a dry ESP unit. The lignite coal-fired electricity-generating unit with an amine post-combustion capture system consists of the following: (i) unit 3 at the BDPS, which generates 150 MW and is a tangentially fired subcritical boiler, (ii) a dry ESP unit, (iii) a wet FGD unit, and (iv) a CO_2 capture and compression unit. The oxy-fuel combustion CO_2 capture system consists of the following: (i) an air separation unit (ASU) for cryogenic distillation, which is often commercially used for air separation, (ii) unit 3 at the BDPS, which generates 150 MW and is a tangentially fired subcritical boiler, (iii) a dry ESP unit, (iv) a wet FGD unit, and (v) a CO_2 purification and compression unit.

The oxy-fuel combustion CO_2 capture technology model is described in [6]. The post-combustion CO_2 capture technology model is presented in [8].

4.2. System boundary

The studied system is located at the BDPS unit 3 in Estevan, Saskatchewan, Canada. From this location, the emissions of heavy metals are predicted to occur in a circular pattern of 10 degrees increments with 25 points of 100 m on each increment. Each direction has 25 distances starting from 100 m and increases every 100 m. The location of the stack at the BDPS unit 3 is set as an origin of the emissions and designated as (0.0, 0.0).

4.3. Modeling air dispersion and risk

Since the objective of this study is to evaluate the risk to humans posed by the conventional coal-fired power plant, the post-combustion, and oxygen-based combustion systems specific to Saskatchewan, Canada, the evaluation was conducted using methodologies for assessing air pollution dispersion, cancer, and non-cancer risks. Two options were considered for implementing the air pollution dispersion methodology: AERMOD and CALPUFF. AERMOD is a steady-state Gaussian plume dispersion model, which is designed to predict near-field (<50 km) impacts [25]. The model aims to estimate and calculate how the pollutions, which are emitted from a source, can disperse in the atmosphere and travel across a receptor grid [26]. By contrast, CALPUFF is a non-steady-state meteorological and air quality modeling system, which can be applied to measure air quality from tens to hundreds of kilometers [27, 28]. The model consists of preprocessing and post-processing programs that can be categorized into three main components: (1) a meteorological model, (2) an air dispersion model, and (3) post-processing packages for the meteorological, concentration, and deposition data output [29]. Both AERMOD and CALPUFF were developed by the US EPA. Since the Government of

Saskatchewan provides the meteorological data specific to Estevan required in the AERMOD model, and AERMOD has been widely used for predicting near-field impacts of chemical pollutants, the AERMOD model is suitable because this study aims to evaluate the risks to health that people who live near the power station face.

Due to the limited available data on the heavy metals, the equations for calculating cancer and non-cancer risks from [30, 31] were chosen as the most appropriate tools for conducting the risk analysis.

4.3.1. Modeling air dispersion

As previously stated, AERMOD is a steady-state Gaussian plume dispersion model, which is designed to predict near-field (or less than 50 km)impacts in both simple and complex terrains as shown in **Figure 2** [25, 32]. The model recognizes the manner in which the pollutants emitted from a source are dispersed in the atmosphere and travel across a receptor grid [26].

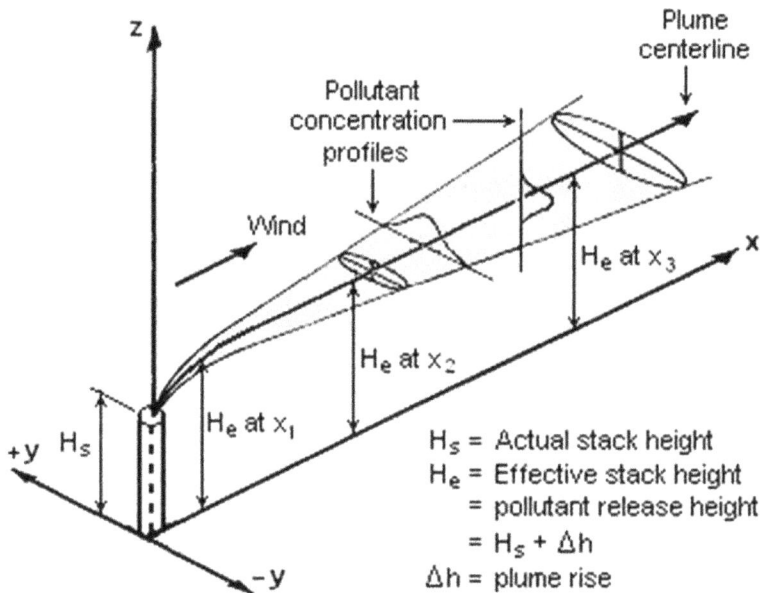

Figure 2. Steady-state Gaussian plume dispersion model in AERMOD [32].

The main data requirements for AERMOD include AERMET, or meteorological data in Estevan, emission rates released from the selected stack, stack height, exit temperature and velocity of the selected emission, and inside stack diameter. The sources of data consist of (i) the meteorological dataset specific to Estevan required in the AERMOD model, which has been provided by the Government of Saskatchewan (www.environment.gov.sk.ca); (ii) the stack data for the "no capture" and "post-combustion" scenario provided by the Saskatche-wan Power Corporation (SaskPower) and the dataset of the oxy-fuel combustion generated using the IECM software version 8.0.2 (Trademark of Carnegie Mellon University, USA), and

(iii) the emission rates from the power plant obtained from the LCA studies of a conventional coal-fired power plant, a post-combustion, and an oxy-fuel combustion CO_2 capture processes [6–8]. The meteorological data from years 2003–2007 were used for the AERMOD modeling due to the limitations in available data. The stack data and emission rates are summarized in **Table 3**.

Coal-fired power plant	BDPS unit 3 (without CO_2 capture)	BDPS unit 3 (with oxy-fuel combustion CO_2 capture)	BDPS unit 3 (with post-combustion CO_2 capture)
Stack height (m)	91.44	91.44	91.44
Stack diameter (m)	4.27	4.27	4.27
Exit gas velocity (m/sec)	18.1	15.09	18.1
Exhaust gas temperature (K)	436.15	310.93	436.15
Mercury (Hg)	0.001902781	–	0.001675417
Antimony (Sb)	0.000170833	1.70833E−06	3.41667E−06
Arsenic (As)	0.002041667	0.000018375	0.00003675
Barium (Ba)	0.000541667	5.41667E−06	1.08333E−05
Beryllium (Be)	6.66667E−05	3.33333E−07	6.66667E−07
Cadmium (Cd)	0.000170833	3.41667E−06	6.83333E−06
Chromium (Cr)	0.002458333	0.0000295	0.000059
Cobalt (Co)	0.0002875	0.000002875	0.00000575
Copper (Cu)	0.000958333	2.97083E−05	5.94167E−05
Lead (Pb)	0.00125	0.0000125	0.000025
Manganese (Mn)	0.000179167	1.25417E−06	2.50833E−06
Molybdenum (Mo)	0.001583333	1.58333E−05	3.16667E−05
Nickel (Ni)	0.002416667	3.38333E−05	6.76667E−05
Selenium (Se)	0.017083333	0.0007175	0.001435
Vanadium (V)	0.003666667	2.93333E−05	5.86667E−05

Table 3. Stack features and emission rates.

A comparison of the three scenarios revealed that the higher temperatures, which cause more atmospheric lift, occur with the stacks in the "no capture" and the "post-combustion capture" scenarios. However, the flow velocity in the "post-combustion capture" scenario should have been slightly lowered because of the pressure drop in the unit processes. This study used the same flow velocity both in the "no capture" and the "post-combustion capture" scenarios because this study has adopted the data on the exhaust gas velocity and temperature from SaskPower, which was the only source of data available. The "oxy-fuel combustion" scenario showed lower exhaust gas velocity and temperatures due to the recycling of the flue gas and the CO_2 compression and purification unit. The data on exit gas velocity was obtained from the SaskPower Web site for the "no capture" and "post-combustion" scenarios, while the oxy-fuel combustion data were results taken from IECM modeling.

4.3.2. Analysis of cancer and non-cancer risks analysis

The risk calculation involves an estimation of the cancer and non-cancer risks related to heavy metals, which can become inhaled contaminants. The emission data for the "no capture" and the two "capture" scenarios are taken from the LCI results in [6–8]. Based on the data, the emission concentrations on the ground were generated using AERMOD, and then, the data were used for evaluating the cancer and non-cancer risks. The equations recommended for estimating cancer and non-cancer risks are taken from [30, 31].

4.3.2.1. Long-term cancer risk

While cancer risks can be associated with both inhalation and ingestion, this study only took the risk related to inhalation into consideration. The unacceptable cancer risk is the risk higher than 1,000,000 [9, 33]. In other words, a cancer risk which is higher than 0.000001 will cause carcinogenic effects, which is an undesirable outcome. The unit risk factor (URF) data were taken from the toxicity values for inhalation exposure shown on the New Jersey Department of Environmental Protection Web site (www.nj.gov). The cancer risk via the inhalation pathway can be calculated with the following equation:

$$\text{Cancer risk} = EC * URF \tag{4.1}$$

where EC = Exposure air concentration ($\mu g/m^3$) and URF = Unit risk factor ($\mu g/m^3$)$^{-1}$.

4.3.2.2. Long- and short-terms non-cancer risk

The exposure to non-cancer risk due to direct inhalation can be estimated using the hazard quotient (HQ) approach, which involves a ratio for estimating chronic dose/exposure level to the reference concentration (RfC), an estimated daily concentration of emissions in the air [30, 34]. There are two main types of RfC values associated with long-term and short-term effects. The RfC data were taken from the toxicity values for inhalation exposure shown on the New Jersey Department of Environmental Protection Web site (www.nj.gov). HQ values equal to or less than one are referred to as having little or no adverse effect [34]. By contrast, a HQ value that exceeds one implies that the emissions have reached a level of concern [35]. However, since the HQ is not a probability of risk, it does not matter how large the HQ value is, only whether or not the HQ value exceeds one [34]. For example, a quotient of 0.01 does not mean that there is a one in a hundred chance that the effect will occur. The HQ value is calculated using the following equation.

$$HQ = EC / RfC \tag{4.2}$$

where HQ = Hazard quotient (unitless), EC = Exposure air concentration ($\mu g/m^3$), and RfC = Reference concentration ($\mu g/m^3$).

5. Results

5.1. Results from AERMOD

The study examined the air dispersion modeling of the "no capture" and the two "capture" scenarios. For cancer and non-cancer risks, the maximum 24-hour and 1-hour average concentration values of heavy metals were used for long-term and short-term exposures, respectively. The maximum 24-hour concentration values generated from AERMOD of the "no capture," "post-combustion CO_2 capture," and "oxy-fuel combustion CO_2 capture" scenarios are shown in **Table 4**. For short-term effects, the maximum 1-hour concentration values generated from AERMOD of the "no capture," "post-combustion CO_2 capture," and "oxy-fuel combustion CO_2 capture" scenarios are shown in **Table 5**. It can be seen from the two tables that the maximum 24-hour and 1-hour average concentrations of the heavy metals of the "no capture" scenario, respectively, show the highest concentrations compared to the other two scenarios. This shows that when the CO_2 capture technologies are applied, lower concentrations of Hg and heavy metals will be emitted into the air. These emissions are captured by the pollution control units provided in the CO_2 capture technologies, and distribution in the atmosphere is controlled by parameters such as the stack height, exhaust gas temperature, and exit gas velocity, as shown in **Table 3**.

Substances	Concentrations		
	No capture	Oxy-fuel combustion	Post-combustion
Hg	4.72E–02	0	4.15E–02
As	5.06E–02	1.08E–03	9.1E–04
Ba	1.34E–02	3.2E–04	2.7E–04
Be	1.65E–03	2.0E–05	2.0E–05
Cd	4.24E–03	2.0E–04	1.7E–04
Cr	6.1E–02	1.73E–03	1.46E–03
Co	7.14E–03	1.7E–04	1.4E–04
Cu	2.37E–02	1.74E–03	1.47E–03
Pb	3.10E–02	7.3E–04	6.2E–04
Ni	5.99E–02	1.99E–03	1.68E–03
Se	4.23E–01	4.21E–02	3.56E–02
V	9.1E–02	1.72E–03	1.46E–03

Table 4. The maximum 24-hour average concentrations of the heavy metals of the "no capture" and the two "capture" scenarios in 2003–2007 ($\mu g/m^3$).

Substances	Concentrations		
	No capture	Oxy-fuel combustion	Post-combustion
Hg	4.16E−01	0	3.66E−01
As	4.47E−01	8.43E−03	8.05E−03
Ba	1.18E−01	2.48E−03	2.37E−03
Be	1.46E−02	1.5E−04	1.5E−04
Cd	3.74E−02	1.57E−03	1.5E−03
Cr	5.38E−01	1.35E−02	1.29E−02
Co	6.29E−02	1.32−03	1.26E−03
Cu	2.09E−01	1.36E−02	1.3E−02
Pb	2.73E−01	5.73E−03	5.47E−03
Ni	5.29E−01	1.55E−02	1.48E−02
Se	3.74	3.29E−01	3.14E−01
V	8.02E−01	1.34E−02	1.28E−02

Table 5. The maximum 1-hour average concentrations of the heavy metals of the "no capture" and the two "capture" scenarios in 2003–2007 ($\mu g/m^3$).

The oxy-fuel combustion system gives out less emission at a lower flow velocity, so the emissions fall on the ground closer to the stack and there are less emissions further away. By contrast, the post-combustion system gives out higher emissions at a higher velocity, which enables the emissions to travel further away; the higher temperature of the flue gas also causes atmospheric lift of the emissions. As a result, the emissions are more evenly distributed over a wider area further away from the stack, and their concentrations are lower.

5.2. Results from cancer and non-cancer risks related to heavy metals

The missing inhalation URF and RfC values limit the calculations of cancer and non-cancer risks for some metals. Cancer and non-cancer risk results are shown in **Table 6** and **Table 7**, respectively. **Tables 6** indicates that the emissions from the stack in each of the three scenarios pose cancer risks of less than one chance in a million (1×10^{-6}). However, there are two emissions, which include As and Cr, from the "no capture" scenario that pose cancer risks due to inhalation with a chance greater than 1×10^{-6}. In terms of non-cancer risks, the inhalation exposures are estimated by the HQ value, a ratio to estimate chronic dose/exposure level to RfC, an estimated daily concentration of emissions in air. The results shown in **Table 7** display that all HQ values are less than one. When the HQ values are less than one, this indicates that pollutant concentrations from the three stacks are unlikely to correlate with any non-cancer-related health concerns.

Substances	Inhalation unit risk factor (URF) $(\mu g/m^3)^{-1}$	Cancer risk		
		No capture	Oxy-fuel combustion	Post-combustion
Hg	–	–	–	–
As	4.3E–03	1.45E–06	3.09E–08	2.61E–08
Ba	–	–	–	–
Be	2.4-E03	2.64E–08	3.20E–10	3.20E–10
Cd	4.2E–03	1.18E–07	5.6E–09	4.76E–09
Cr	1.2E–02	4.88E–06	1.38E–07	1.16E–07
Co	9E–03	4.28E–07	1.02E–08	8.4E–09
Cu	–	–	–	–
Pb	1.2E–05	2.48E–09	5.84E–11	4.96E–11
Ni	–	–	–	–
Se	–	–	–	–
V	–	–	–	–

Table 6. Cancer risks of heavy metals.

Substances	Long term				Short term			
	RfC ($\mu g/m^3$)	Non-cancer risk			RfC ($\mu g/m^3$)	Non-cancer risk		
		No capture	Oxy-fuel combustion	Post-combustion		No capture	Oxy-fuel combustion	Post-combustion
Hg	0.3	1.05E–03	0	9.24E–04	–	–	–	–
As	0.015	2.25E–02	4.8E–04	4.04E–04	0.2	1.49E–02	2.81E–04	2.68E–04
Ba	–	–	–	–	0.5	1.58E–03	3.31E–05	3.16E–05
Be	0.4	5.5E–04	6.67E–06	6.67E–06	–	–	–	–
Cd	0.02	1.41E–03	6.67E–05	5.56E–05	–	–	–	–
Cr	–	–	–	–	–	–	–	–
Co	0.006	7.93E–03	1.88E–04	1.55E–04	–	–	–	–
Cu	–	–	–	–	100	1.39E–05	9.09E–07	8.67E–07
Pb	–	–	–	–	0.1	1.82E–02	3.82E–04	3.65E–04
Ni	0.05	7.99E–03	2.65E–04	2.24E–04	6	5.87E–04	1.72E–05	1.65E–05
Se	20	1.41E–04	1.40E–05	1.18E–05	–	–	–	–
V	0.1	6.06E–03	1.14E–04	9.73E–05	–	–	–	–

Table 7. Long- and short-term inhalation exposures of heavy metals.

6. Discussion

The carbon capture technology is one of the most widely discussed solutions for cutting GHG emissions which are mostly generated from electrical generation that uses fossil fuels (e.g., oil, coal, and natural gas, which are regarded as the world's primary source of energy). According to [36], fossil fuels will be continuously used to supply energy globally for at least the next few decades, especially with the recent development of shale gas in many regions of the world. In this scenario, without a proper control technique, the CO_2 atmospheric emissions will continue to increase and pose an even more serious threat to people and the environment. To cope with this problem, the adoption and use of an effective CO_2 capture technology have become an important approach in ensuring the reduction CO_2 emissions. Consequently, it is important to conduct risk assessment to ensure safety of the carbon capture technology. Understanding those risks can support the formulation of standards and regulatory frameworks required for large-scale application of the carbon capture technology [5]. Greater emissions of carbon dioxide poses hazards to human health because inhaling concentrations of CO_2 emissions around 3–5% will pose risks to human health [37]. Inhaling concentration higher than 15% can be fatal. The health, safety, and environmental (HSE) risk of the fossil-fuel-based electrical generation system can be determined to a large extent by both the total amount of CO_2 lost and the maximum rate of CO_2 lost in the system [2]. The health-related damage associated with emissions from coal-fired electricity-generating plants can vary, depending on a number of factors including the facilities, the function of the plant, the site, and population characteristics [38].

Different studies focus on different kinds of risks associated with the process of carbon capture such as (1) cancer and non-cancer risks; (2) population exposure per unit of emissions, which is associated with atmospheric condition, the population size, and their proximities to the emissions; (3) social and mental impacts; and (4) accidents and deaths [9, 14, 15, 39–42]. According to [9], among the emissions from coal-fired electricity-generating plants, As and Cr were the main contributors to cancer risks, and HCl, Mn, HF, and Hg contributed to the non-cancer risks. The coal combustion process can also release many toxic elements, which include As, Hg, Cd, Pb, Se, and Zn, and among these, Hg is of the most concern [15]. According to [43], the population in Estevan has an exceptionally high rate of asthma. In [44], the study compares the human health risks associated with SO_2, NO_2, and $PM_{2.5}$ of the oxy-fuel carbon dioxide capture with those from the post-combustion CO_2 capture technology, and the study reveals that the oxy-fuel system posed fewer human health risks because this technology captures more emissions. In [44], the study fills the gap in research because none of the past studies emphasize the human health impacts due to heavy metals associated with the BDPS in Estevan, Saskatchewan, Canada. This study produces useful data on human health risk and help decision makers quantify the impact of different CO_2 capture technologies. From a practical perspective, the study provides support for efforts aimed at improving the air quality in the Estevan region.

7. Conclusion

Since the coal-fired electricity generation plant is widely regarded as a significant source of air pollution, the adoption of the carbon capture technology is a potential solution for reducing emissions. However, the carbon capture technology requires additional energy for its operation which results in lowering the overall efficiency of the electricity-generating plant. More fossil fuel per unit of electricity generated is needed to compensate for the lost capacity, but the higher requirement also necessitates a higher level of emissions and resource consumption. Since safety of the carbon capture technology is an important public concern, a risk analysis of the carbon capture technology was conducted. While risk is normally defined as the potential of an unwanted negative consequence or event [17], risk analysis is a tool used to form, structure, and collect information to identify existing hazardous situations and report potential problems or the type and level of the environmental health and safety risk [36].

This study focuses on examining the health impacts of the conventional coal-fired generation station without CO_2 capture, with post-combustion and oxy-fuel combustion CO_2 capture technologies. The study analyzed the cancer and non-cancer risks to human health based on the data of air pollutants from heavy metals obtained from the LCA models [6–8]. The risks associated with these pollutants are calculated for the three CO_2 capture scenarios of (i) "no capture," (ii) "post-combustion CO_2 capture," and (iii) "oxy-fuel combustion CO_2 capture."

7.1. Summary of air dispersion modeling

The maximum 24-hour and 1-hour average concentration values of Hg and heavy metals are used for assessing the long-term and short-term exposures, respectively. The results show that, in the "no capture" scenario, the maximum 24-hour and 1-hour average concentrations of the Hg and heavy metals, respectively, show the highest concentrations compared to the two "capture" scenarios. This shows that these emissions are captured by the pollution control units of the CO_2 capture technologies and the less concentrated Hg and heavy metals consequently will be emitted into the air. The air dispersion modeling, which generates emission concentrations, depends not only on the amount of emissions but also on other parameters such as the stack height, exhaust gas temperature, and exit gas velocity. Compared to the post-combustion system, the oxy-fuel combustion system gives out less emission at a lower flow velocity, so the emissions fall on the ground closer to the stack. As a result, there are less emissions further away.

7.2. Summary of risk analysis

The analysis results shown in **Table 6** indicate that the emissions from the three stacks generally posed cancer risks of less than one chance in a million (1×10^{-6}). However, there are emissions from two elements in the "no capture" scenario that pose cancer risks of more than 1×10^{-6}; As and Cr are the primary contributors to these risks. In terms of non-cancer risks, the results show that all HQ values are less than one. This indicates that the pollutant concentration from the three stacks will not cause any non-cancer health issues.

A limitation in the cancer and non-cancer risks calculation is that data on URF and RfC associated with some types of heavy metals are not available. In future studies, this limitation can be addressed. Generally, it can be concluded that for electricity generation with carbon capture, even though there are increases in adverse health impacts associated with soil and water pollution, the broad distribution of health impacts associated with atmospheric pollutants is significantly reduced. We believe the benefits to human health outweigh the negative of increased emissions.

Acknowledgements

We would like to acknowledge the financial support to the first author from the Networks of Centres of Excellence of Canada–Carbon Management Canada (CMC–NCE), the Government of Saskatchewan, and the Faculty of Graduate Studies and Research of University of Regina. We are also grateful for the financial support from the Canada Research Chair Program to the research project.

Author details

Jarotwan Koiwanit[1], Anastassia Manuilova[2], Christine Chan[1*], Malcolm Wilson[2] and Paitoon Tontiwachwuthikul[1]

*Address all correspondence to: chanchristine888@gmail.com

1 Faculty of Engineering and Applied Science, University of Regina, Saskatchewan, Canada

2 ArticCan Energy Services, Regina, Saskatchewan, Canada

References

[1] IPCC. Summary for policymakers in climate change 2014: Impacts, adaptation, and vulnerability. New York, USA: Intergovernmental Panel on Climate Change; 2014.

[2] Gerstenberger M, Nicol A, Stenhouse M, Berryman K, Stirling M, Webb T, Smith W. Modularised logic tree risk assessment method for carbon capture and storage projects. Energy Procedia. 2009; 1(1): 2495–2502.

[3] Trabucchi C, Donlan M, Wade S. A multi-disciplinary framework to monetize financial consequences arising from CCS projects and motivate effective financial responsibility. International Journal of Greenhouse Gas Control. 2010; 4(2): 388–395.

[4]   Statistics Canada. Census agglomeration of Estevan, Saskatchewan. 2015 [Internet]. Available from: https://www12.statcan.gc.ca/census-recensement/2011/as-sa/fogs-spg/Facts-cma-eng.cfm?LANG=Eng&GK=CMA&GC=750. Accessed on 30 August 2015.

[5]   Damen K, Faaij A, Turkenburg W. Health, safety and environmental risks of underground CO_2 sequestration. (No. NWS-E-2003-30). Netherlands: Copernicus Institute for Sustainable Development and Innovation; 2003.

[6]   Koiwanit J, Manuilova A, Chan C, Wilson M, Tontiwachwuthikul P. A life cycle assessment study of a hypothetical Canadian oxy-fuel combustion carbon dioxide capture process. International Journal of Greenhouse Gas Control. 2014; 28: 257–274.

[7]   Koiwanit J, Piewkhaow L, Zhou Q, Manuilova A, Chan C W, Wilson M, Tontiwachwuthikul P. A life cycle assessment study of a Canadian post-combustion carbon dioxide capture process system. The International Journal of Life Cycle Assessment. 2014; 19(2): 357–369.

[8]   Manuilova A. Evaluation of environmental performance of carbon capture and storage project in Canada using life cycle assessment methodology [thesis]. Regina: University of Regina; 2011.

[9]   French C, Peters W, Maxwell B, Rice G, Colli A, Bullock R, Cole J, Heath E, Turner J, Hetes B, Brown D C, Goldin D, Behling H, Loomis D, Nelson C. Assessment of health risks due to hazardous air pollutant emissions from electric utilities. Drug and Chemical Toxicology. 1997; 20(4): 375–386.

[10] Wu Y, Rahmaningrum D, Lai Y, Tu L, Lee S, Wang L, Chang-chien G. (Mercury emissions from a coal-fired power plant and their impact on the nearby environment. Aerosol and Air Quality Research. 2012; 12: 643–650.

[11] Srivastava R K, Hutson N, Martin B, Princiotta F, Staudt J. Control of mercury emissions from coal-fired electric utility boilers. Environmental Science & Technology. 2006; 40(5): 1385–1393.

[12] Lee S, Keener T C. Dispersion modeling of mercury emissions from coal-fired power plants at Coshocton and Manchester, Ohio. Ohio Journal of Science. 2008; 108(4): 65–69.

[13] Tudose T, Moldovan F. Characteristics of heavy rainfall parameters in the north-western Romania. Aerul şi Apa: Componente ale Mediului. 2011; 91–98.

[14] Mokhtar M M, Hassim M H, Taiba M R, Lim S Z, Sahani M. Health risk assessment in coal-fired power plant in Malaysia. In: Proceedings of the 6th International Conference on Process Systems Engineering (PSE ASIA); 2013. p. 147–152.

[15] Castleden W M, Shearman D, Crisp G, Finch P. The mining and burning of coal: effects on health and the environment. The Medical Journal of Australia. 2011; 195(6): 333–335.

[16] CDC. Fourth national report on human exposure to environmental chemicals. Atlanta, GA, USA: Centers for Disease Control and Prevention; 2009.

[17] Elizabeth L A, Roy E A. Risk assessment and indoor air quality. New York, USA: Lewis Publishers; 1998.

[18] Hu H. (2002). Human health and heavy metals exposure. In: McCally M, editor. Life support: The environment and human health; MIT Press. Cambridge, Massachusetts. 2002. p. 65–82.

[19] Keating M. Cradle to grave: the environmental impacts from coal. Boston, MA, USA: Clean Air Task Force; 2001.

[20] NH DES. Copper: health information summary. Concord, NH, USA: New Hampshire Department of Environmental Services; 2013.

[21] US EPA. Selenium compounds. 2000. [Internet]. Available from:http://www.epa.gov/ ttnatw01/hlthef/selenium.html. Accessed on 15 Auguest 2013.

[22] Baumann H, Tillman A. The hitch hiker's guide to LCA. An orientation in life cycle assessment methodology and application. United States of America: Studentlitteratur; 2004.

[23] Beacon news group Canada. SaskPower launches world's largest carbon capture project. Edmonton Beacon; 2014.

[24] Environment Canada. Forecast regions - Saskatchewan. 2013. [Internet]. Available from: http://www.ec.gc.ca/meteo-weather/default.asp?lang=En&n=CE708E88-1. Accessed on 21 November 2014.

[25] US EPA. Summary of public comments: 10th conference on air quality modeling. (No. EPA - HQ - OAQ - 2012 - 0056). U.S.: US EPA, Washington, DC; 2012.

[26] Heckel P F, LeMasters G K. The use of AERMOD air pollution dispersion models to estimate residential ambient concentrations of elemental mercury. Water, Air, & Soil Pollution. 2011; 219(1–4): 377–388.

[27] Hoeksema G, Onder K, Unrau G. A comparison of Aermod and Calpuff models for regulatory dispersion modelling in the alberta oil sands region. Air and Waste Management Association Annual Meeting Conference and Exhibition. 2011; 3: 2035–2044.

[28] US EPA. CALPUFF modeling system. 2013. [Internet]. Available from: http://www.epa.gov/scram001/dispersion_prefrec.htm. Accessed on 4 April 2014.

[29] Scire J S, Strimaitis D G, Yamartino R J. A user's guide for the CALPUFF dispersion model. Concord, MA, USA: Earth Tech, Inc; 2000.

[30] US EPA. Chapter 7: Characterizing risk and hazard. Human health risk assessment protocol (pp. 7-1-7-15). 2005; US EPA, Washington, DC.

[31] US EPA. Risk assessment guidance for superfund. volume I: Human health evaluation manual: (Part F, supplemental guidance for inhalation risk assessment). (No. EPA-540-R-070-002). Washington, DC: US EPA; 2009.

[32] Beychok M. Air pollution dispersion modeling; 2012. Available from: http://www.eoearth.org/view/article/169860/ [Accessed 2014-11-21].

[33] ATSDR. Public health assessment guidance manual (2005 update): Appendix F: Derivation of comparison values [Internet]; 2005. Available from: http://www.atsdr.cdc.gov/hac/PHAManual/appf.html [Accessed: 2014-12-15].

[34] US EPA. Ecological risk assessment step 2 [Internet]; 2011. Available from: http://www.epa.gov/R5Super/ecology/erasteps/erastep2.html [Accessed: 2014-12-15].

[35] Kincaid L E, Davis G A, Meline J. Cleaner technologies substitutes assessment – A methodology and resource guide. Washington, DC: US EPA; 1997.

[36] Suebsiri J. A model of carbon capture and storage with demonstration of global warming potential and fossil fuel resource use efficiency [thesis]. Regina: University of Regina; 2010.

[37] Lilliestam J, Bielicki J M, Patt A G. Comparing carbon capture and storage (CCS) with concentrating solar power (CSP): Potentials, costs, risks, and barriers. Energy Policy. 2012; 47: 447–455.

[38] Levy J I, Baxter L K, Schwartz J. Uncertainty and variability in health-related damages from coal-fired power plants in the United States. Risk Analysis. 2009; 29(7): 1000–1014.

[39] Gjernes E, Helgesen L I, Maree Y. Health and environmental impact of amine based post combustion CO_2 capture. Energy Procedia. 2013; 37: 735–742.

[40] Nielsen C J, Herrmann H, Weller C. Atmospheric chemistry and environmental impact of the use of amines in carbon capture and storage (CCS). Chemical Society Reviews, 2012; 41(19): 6684–6704.

[41] Preiss P, Roos J, Friedrich R. Assessment of health impacts of coal fired power stations in Germany. Stuttgart, Germany: Institute for Energy Economics and the Rational Use of Energy (IER); 2013.

[42] Senior C L, Morris W, Lewandowski T A. Emissions and risks associated with oxyfuel combustion: state of the science and critical data gaps. Journal of the Air & Waste Management Association. 2013; 63(7): 832–843.

[43] The University of Saskatchewan Airways Research Group. The impact of airborne environmental contaminants on respiratory public health in Saskatchewan; 2012. Unpublished manuscript.

[44] Koiwanit J. Evaluation of environmental performance of hypothetical Canadian oxy-fuel combustion carbon capture with risk and cost analyses [thesis]. Regina: University of Regina; 2015.

Livestock as Sources of Greenhouse Gases and Its Significance to Climate Change

Veerasamy Sejian, Raghavendra Bhatta, Pradeep Kumar Malik, Bagath Madiajagan, Yaqoub Ali Saif Al-Hosni, Megan Sullivan and John B. Gaughan

Additional information is available at the end of the chapter

Abstract

This chapter outlines the role of livestock in the production of greenhouse gases (GHGs) that contributes to climate change. Livestock contribute both directly and indirectly to climate change through the emissions of GHGs such as carbon dioxide (CO_2), methane (CH_4), and nitrous oxide (N_2O). As animal production systems are vulnerable to climate change and are large contributors to potential global warming, it is vital to understand in detail enteric CH_4 emission and manure management in different livestock species. Methane emissions from livestock are estimated to be approximately 2.2 billion tonnes of CO_2 equivalents, accounting for about 80% of agricultural CH_4 and 35% of the total anthropogenic CH_4 emissions. Furthermore, the global livestock sector contributes about 75% of the agricultural N_2O emissions. Other sources of GHG emission from livestock and related activities are fossil fuels used for associated farm activities, N_2O emissions from fertilizer use, CH_4 release from the breakdown of fertilizers and from animal manure, and land-use changes for feed production. There are several techniques available to quantify CH_4 emission, and simulation models offer a scope to predict accurately the GHG emission from a livestock enterprise as a whole. Quantifying GHG emission from livestock may pave the way for understanding the role of livestock to climate change and this will help in designing appropriate mitigation strategies to reduce livestock-related GHGs.

Keywords: climate change, enteric methane, GHG, livestock, manure management, modeling

1. Introduction

The Intergovernmental Panel on Climate Change (IPCC), convened by the United Nations, has reported evidence that human activities over the past 50 years have influenced the global climate through the production of GHG [1]. Increasing concentrations of GHGs in the atmosphere have contributed to an increase in the Earth's atmospheric temperature, an occurrence known as global warming [2]. Indeed, average global temperatures have risen considerably, and the IPCC [1] predicts increases of 1.8–3.9°C (3.2–7.1°F) by 2100. With business as usual, Earth's temperature may rise by 1.4–5.8°C by the end of this century, and the scientific community warns of more abrupt climatic change in the future [3].

The livestock sector accounts for 40% of the world's agriculture gross domestic product (GDP). It employs 1.3 billion people and creates livelihoods for one billion of the world's population living in poverty [2]. As animal production systems are vulnerable to climate change and are large contributors to potential global warming through methane (CH_4) and nitrous oxide (N_2O) production, it is vital to understand in detail enteric CH_4 emission and manure management in different livestock species [4]. Before targeting GHG reduction strategies from enteric fermentation and manure management, it is important to understand the mechanisms of enteric CH_4 emission in livestock, the factors influencing such emission. In addition, an understanding of the available prediction models and estimation methodology for quantification of GHGs is essential. A thorough understanding of these will in turn pave way for formulation of effective mitigation strategies for minimizing enteric CH_4 emission in livestock [5].

This chapter will focus on four main areas: (i) livestock's role as a source of GHGs, and the contribution that this makes to climate change; (ii) enteric CH_4 emission and manure management related to CH_4 and N_2O as primary sources of GHGs related to livestock activities; (iii) the methodologies used to quantify enteric emission; and (iv) modeling of GHGs in livestock farms as important step towards finding solution for livestock-related climate change.

2. Livestock and climate change from global food security perspectives

FAO estimated that 1526 million cattle and buffaloes and 1777 million small ruminants are being maintained globally. The population of cattle and buffaloes and small ruminants is expected to be 2.6 and 2.7 billion, respectively, by the year 2050. Furthermore, livestock are an integral element of agriculture that supports the livelihood of more than 1 billion people across the globe. This sector satisfies more than 13% of the caloric and 28% of the protein requirements of people worldwide. The global demand for milk, meat, and eggs is expected to increase by 30%, 60%, and 80%, respectively, by the year 2050 in comparison to the 1990 demand. This increased requirement will be fulfilled either by increasing the livestock numbers or through intensifying the productivity of existing stock.

Climate change is seen as a major threat to the survival of many species, ecosystems, and the sustainability of livestock production systems in many parts of the world [6]. The growing human population will almost doubled the global requirement for livestock products by 2050. It is during the same period adverse changes in the climate are also expected. Recent industrial developments have curtailed the land used for agricultural activities, considerably threatening food security in both developed and developing countries. Hence, livestock production has a key role to play in bringing food security to these countries. We need high-quality research in animal science to meet the increasing demand for livestock products in the changing climate scenario [7].

3. Livestock as source of greenhouse gas (GHGs)

Livestock contributes both directly and indirectly to climate change through the emissions of GHGs such as CO_2, CH_4, and N_2O [8]. Globally, the sector contributes 18% (7.1 billion tonnes CO_2 equivalent) of global GHG emissions. Although it accounts for only 9% of global CO_2, it generates 65% of human-related N_2O and 35% of CH_4, which has 310 times and 23 times the global warming potential (GWP) of CO_2, respectively [9] (Fig. 1).

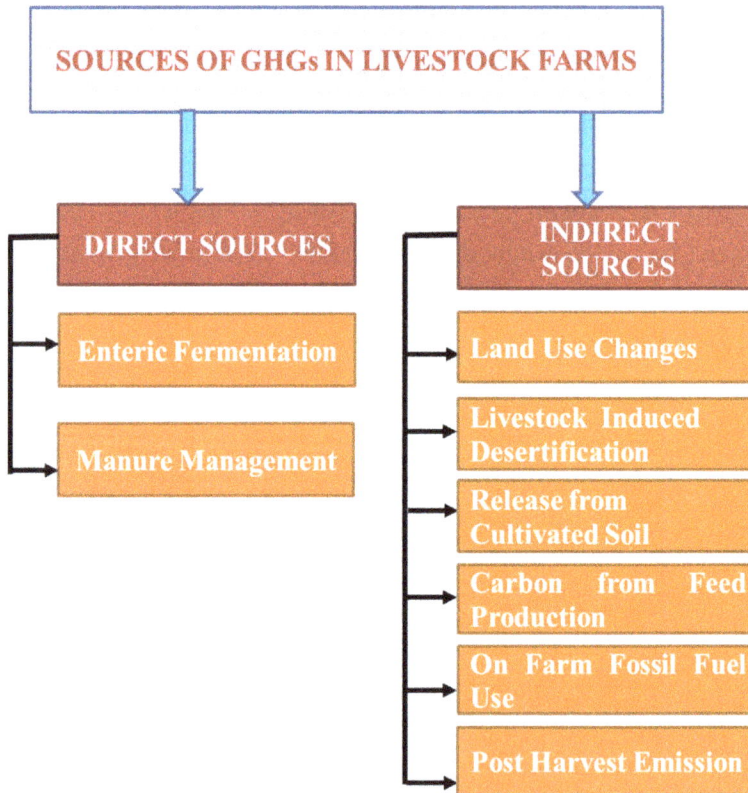

Figure 1. Different sources of GHGs from livestock sector.

There are two sources of GHG emissions from livestock: (a) enteric fermentation where specific microbes residing in the rumen produce CH_4 as a by-product during digestion and (b) anaerobic fermentation of livestock manure producing CH_4 and denitrification and denitrification of manure producing N_2O. Methane production appears to be a major issue and largely arises from natural anaerobic ecosystems, and fermentative digestion in ruminant animal [10]. Much of the global GHG emissions currently arise from enteric fermentation and manure from grazing animals. The development of management strategies to mitigate CH_4 emissions from ruminant livestock is possible and desirable. Carbon dioxide (CO_2) are also produced in livestock farms and are primarily associated with fossil fuel burning during operation of farm machineries in the process of fertilizer production, processing and transportation of refrigerated products, deforestation, desertification, and release of carbon from cultivated soils. Enhanced utilization of dietary "C" will improve energy utilization and feed efficiency hence animal productivity, decrease overall CH_4 emissions, and thereby reduce the contribution of ruminant livestock to the global CH_4 inventory.

Ruminant animals, such as cattle, sheep, buffaloes, and goats, are unique due to their special digestive systems, which can convert plant materials that are indigestible by humans into nutritious food. In addition to food, these animals also produce hides and fibers that are utilized by humans. This same helpful digestive system, however, produces CH_4, a potent GHG that can contribute to global climate change. Livestock production systems can also emit other GHGs such as N_2O and CO_2. The most important GHGs are CO_2, CH_4, and N_2O, all of which have increased in the last 150 years and have different global warming potential. According to Sejian et al. [11], the warming potential of CO_2, CH_4, and N_2O are 1, 25, and 310, respectively. Taking into account the entire livestock commodity chain – from land use and feed production, to livestock farming and waste management, to product processing and transportation – about 18% of total anthropogenic GHG emissions can be attributed to the livestock sector [2].

Livestock production is the largest global source of CH_4 and N_2O – two particularly potent GHGs [12, 13]. The principal sources of N_2O are manure and fertilizers used in the production of feed. The biggest source of CH_4 is from enteric fermentation. The rising demand for livestock products therefore translates into rising emissions of CH_4 and N_2O. According to one study, if current dietary trends (increasing global consumption of animal products) were to continue, emissions of CH_4 and N_2O would more than double by 2055 from 1995 levels [14].

3.1. Enteric methane emission

Worldwide livestock emits around 7.1 Gt CO_2-eq GHGs per year, which accounts for 15% of the human induced GHGs emissions. Additionally, 5.7 Gt CO_2-eq GHGs is also emitted from the ruminant supply chain wherein cattle, buffaloes, and small ruminant production contribute 81%, 11%, and 8%, respectively. Methane emissions from livestock have two sources, one from enteric fermentation and another from excrement. Enteric fermentation in ruminants annually contributes ~90 Tg CH_4 to the atmospheric pool, while ~25 Tg comes from the excrement. Apart from the role of enteric CH_4 in global warming, its emission from the animal system lead to a loss of biological energy (6–12% of intake), which otherwise would have been utilized by the host animal for various productive functions. Reducing the loss of energy in the form of enteric CH_4 is crucial, especially in developing countries like India where feed and fodder availability is already in short supply.

3.1.1. Indian livestock and enteric methane emission

India has approximately 512 million livestock (19th Livestock census, Government of India). Of the total livestock population, about 60% are cattle and buffaloes, which comparatively emit more enteric CH_4 than any other livestock species. Emissions of enteric CH_4 can be elevated when these species are fed fibrous feeds. Estimations of enteric CH_4 emissions from Indian livestock have been calculated using different approaches (Table 1). There is a lack of consistency in the published data; some have reported very high emissions from Indian livestock while others have reported much lower emissions [15]. This large variation in predicted enteric CH_4 emission from Indian livestock is attributed to the different approaches used for the calculation of emissions. The average of the published data is in the range of 9–10 Tg per year, which appears to be a realistic value. Methane emissions from excrement in India are low because the disposal system (generally stored as heap in the open environment) does not support the favorable anaerobic conditions required by methanogens. However, in the developed world where excrement is mainly stored in lagoons, manure is a major source of CH_4 emissions.

Source	Base year	Emission (Tg/yr)	Approach
Ahuja [16]	1985	10.40	Default CH_4 emission factors
ALGAS[a] [17]	1990	18.48	IPCC methodology
Singh [18]	1992	9.02	In vitro gas production and dry matter digestibility coefficients
Garg and Shukla [15]	-	7.25	-
EPA[b] [19]	-	10.04	-
Swamy and Bhattacharya [20]	1994	9.0	Methane emission factors
Jha et al. [21]	1994	8.97	IPCC tier II
Chhabra et al. [22]	2003	10.65	GIS approach
Singh et al. [23]	2003	9.10	In vitro gas production, feeding practices in different agro-ecological regions
Patra [24]	2010	14.3	IPCC tier 1

[a]Asia Least-Cost Greenhouse Gas Abatement Strategy.

[b]United States Environmental Protection Agency.

Table 1. Estimates of enteric methane emission from Indian livestock

Based on the IPCC default emission factors, Kamra [25] determined the enteric CH_4 emission from Indian livestock. Buffalo, yak, and mithun contribute a maximum of 55 kg CH_4/head/yr; however, sheep and goat contribute only 5 kg/head/yr. The enteric CH_4 emission from crossbred cattle is much higher than the indigenous cattle (46 vs 25 kg/head/yr). Both cattle and buffaloes aggregately emit more than 90% of the total enteric CH_4 from livestock, while sheep and goat together contribute around 7.70% (Fig. 2). Pig production is the next major emitter contributing 0.57% of the total enteric CH_4 emission from livestock in India. The contribution from other livestock species is negligible.

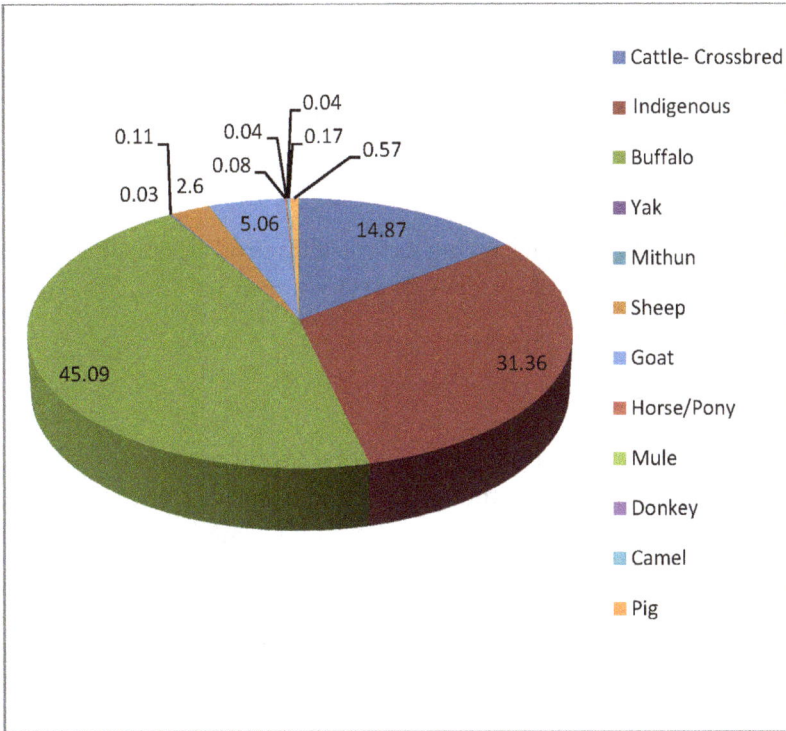

Figure 2. % Species wise enteric methane emission (modified from Kamra [25]).

3.1.2. Why rumen methanogenesis is an obligation

The rumen is the harbor for diverse anaerobic microbe populations that accomplish different functions from degradation of complex carbohydrates to the removal of fermentation metabolites in a syntrophic way [26]. H_2, which is produced in large volumes during enteric fermentation, needs to be removed from the anaerobic vat in order to maintain favorable rumen conditions for both the rumen microbes and host animal. Under normal rumen functioning, metabolic H_2 is used for the reduction of CO_2 to CH_4, which in turn is eructated into the atmosphere via the mouth and nostrils. The microbes of the so-called *archaea* or methanogens are the CH_4 producing machinery inside the rumen. The majority of the rumen methanogens are hydrogenotrophic, which utilize H_2 as a substrate for methanogenesis.

Rumen methanogenesis is a necessary but energy-wasteful process as it corresponds to a significant loss of biological dietary energy (6–12% of intake) in the form of CH_4.

Among the various end products of rumen fermentation, H_2 is a central metabolite where its partial pressure in the rumen determines the extent of methanogenesis and the possible extent of oxidation of feedstuffs [27]. H_2 in the rumen is generally referred as *currency of fermentation* [27]. The removal of H_2 from the rumen is a prerequisite for the continuation of rumen fermentation. However, the methanogens constitute only a small fraction of the rumen microbial community, but they are very crucial in H_2 utilization [28]. Apart from the methanogenesis, other hydrogenotrophic pathways (reductive acetogenesis, sulfate and nitrate reduction) are also present in the rumen, but the extent of H_2 utilization through these pathways is not clear. In order to keep the rumen functional and the animal alive, rumen methanogenesis is the primary and thermodynamically efficient way of metabolic H_2 disposal from the rumen, and that is why it is generally regarded as a necessary but wasteful process.

3.1.3. Enteric methane estimation methodology

Several methods are available for measuring enteric CH_4 production, and the selection of the most appropriate method is based on several factors such as cost, level of accuracy, and experimental design [29, 30].

3.1.3.1. Individual animal techniques

By far, the most suitable method to quantify individual ruminant animal CH_4 measurement is by using respiration chamber, or calorimetry. The respiration chamber models include whole animal chambers, head boxes, or ventilated hoods and face masks. These methods have been effectively used to collect information pertaining to CH_4 emissions in livestock. The predominant use of calorimeters has been in energy balance experiments where CH_4 has been estimated as a part of the procedures followed. Although there are various designs available, open-circuit calorimeter has been the one widely used. There are various designs of calorimeters, but the most common one is the open-circuit calorimeter, in which outside air is circulated around the animal's head, mouth, and nose and expired air is collected for further analysis.

3.1.3.2. Tracer gas techniques

Methane emission from ruminants can be estimated by using the ERUCT technique (Emissions from Ruminants Using a Calibrated Tracer). The tracer can either be isotopic or nonisotopic. Isotopic tracer techniques generally require simple experimental designs and relatively straightforward calculations [31]. Isotopic methods involve the use of (3H-)CH_4 or (14C-)CH_4 and ruminally cannulated animals.

3.1.3.3. Sulphur hexafluoride (SF_6) technique

Nonisotopic tracer techniques are also available for measurement of CH_4 production. Johnson et al. [32] described a technique using SF_6, an inert gas tracer. This method has been widely used in sheep and cattle. Methane emission rates are calculated based on the equation $QCH_4 = QSF_6 \times [CH_4] / [SF_6]$, where QCH_4 is the emission rate of CH_4 in g/day, QSF_6 is the known release rate (g/day) of SF_6 from the permeation tube, and $[CH_4]$ and $[SF_6]$ are the measured concentrations in the canister.

3.1.3.4. In vitro gas production technique (IVGPT)

Various aspects of in vitro gas production test have been reviewed by Getachew et al. [33], and these authors reported that gas measurement were centered on investigations of rumen microbial activities using manometric measurements and concluded that these methods do not have wide acceptability in routine feed evaluation since there was no provision for the mechanical stirring of the sample during incubation. Another *in vitro* automated pressure transducer method for gas production measurement was developed by Wilkins [34], and the method was validated by Blummel and Orskov [35] and Makkar et al. [36]. There are several other gas-measuring techniques such as (i) Hohenheim gas method or Menke's method [37]; (ii) liquid displacement system [38]; (iii) manometric method [39]; (iv) pressure transducer systems: manual [40], computerized [41], and combination of pressure transducer and gas release system [42].

3.2. Livestock manure as an important source of GHGs

In addition to enteric CH_4 production, livestock manure contributes directly and indirectly to GHG gas production via CH_4, N_2O, and CO_2 production. Manure from livestock includes both dung and urine. Manure management plays a key role in amount of CH_4 and N_2O produced and liberated into the environment. The amount of CH_4 produced in solid-state manure management contribute less when compared to liquid state. However, dry anaerobic management system provides suitable environment for N_2O production. The liquid/slurry manure systems provide favorable environments for the growth of the microbes, which in turn enhances the CH_4 gas production. Various factors that affect CH_4 and N_2O production include the amount of manure, the VFA present, the type of feed, the management systems, and the ambient temperature. In addition, the duration of the storage of waste also influences N_2O production,

3.2.1. Methane emission from manure management

Anaerobic digestion processes occur in manure with the help of microbial consortia to produce CH_4 and CO_2 and consists of four phases: (i) hydrolysis of complex organic particulate matter into simpler low molecular weight compounds; (ii) acidogenesis of simpler low molecular weight organic compounds to organic acids and alcohol; (iii) acetogenesis of organic acids and alcohols to H_2, CO_2, acetic acid, and acetate; and (iv) methanogenesis involves the consumption of acids or hydrogen to produce CH_4 and CO_2. The aforementioned four phases are done by four different groups of bacterial consortia, namely, hydrolytic bacteria, acidogenic bacteria, acetogenic bacteria, and methanogenic bacteria, respectively [43]. CH_4 is also emitted from the collection yard, but it is a minor source. The greatest amount of CH_4 is emitted during storage especially in slurry, the reason being the prevalence of complete anaerobic environment. Solid manure also acts as a source of CH_4 emission. CH_4 is emitted immediately after manure application to the field; however, once the O_2 diffuses into manure, it inhibits CH_4 production [44].

3.2.1.1. Factors affecting methane production from manure

There are several factors that affect the CH_4 production from manure, which includes temperature, organic matter present, microbe load, pH, moisture, and type of feed. However, CH_4 emitted from manure depends primarily on (i) the management system such as solid disposal system, liquid disposal systems, e.g., ponds, lagoons, and tanks, which can emit up to 80% of manure-based CH_4 emissions, while solid manure emits little or no CH_4. (ii) Environmental conditions are also important. The higher the temperature and moisture, the more CH_4 produced. (iii) CH_4 emissions also depend on the quantity of the manure produced, which depends on the number of animals housed, the amount of feed the consumed, and the digestibility of the feed. (iv) Manure characteristics depend on the animal type, feed quality, and rumen microbes present in the rumen and digestive tracks. Manure handled in liquid form tends to release more amount of CH_4 when compared to solid or manures thrown into the pasture, which do not decompose anaerobically. High temperatures with neutral pH and high moisture content enhance CH_4 production [45].

3.2.2. Nitrous oxide emission from manure management

Nitrous oxide is produced from manure by nitrification, denitrification, leaching, volatilization, and runoff. Nitrification and denitrification are direct emissions. N_2O is 16 times more potent than CH_4 and 310 times more potent than CO_2 over a 100-year period [46]. N_2O acts as a source of NO in the stratosphere, which indirectly causes depletion of ozone (O_3), increasing UV radiation reaching the Earth's surface [47]. An increase in animal stocking rates and intensive gazing results in the deposition of huge amounts of N via animal excreta (urine + dung); farm management practices that enhance soil organic N mineralization also lead to N_2O production [1, 48, 49].

3.2.2.1. Mechanism of nitrous oxide production

The emission of N_2O from manure occurs directly by both nitrification and denitrification of nitrogen contained in the manure. This emission mainly depends on the N and C content of the manure during various types storage and treatment. The nitrification process strictly needs oxygen, while subsequent denitrification is an anaerobic process.

Manure from livestock mixes with the soil or in the tank, lagoons, etc., where the microbes break down organic N to inorganic NH_4+ through mineralization. In this step, the organic N becomes available for plants and microorganisms. Microorganisms (*Nitrosomonas* genus) can take up NH_4+ and oxidize it to nitrite (NO_2-). In the next step, *Nitrobacter* and *Nitrococcus* oxidize NO_2- to nitrate (NO_3-) by nitrification. This process of oxidation of NH_4+ to NO_3- is known as nitrification, which is also done by other genera like *Nitrosococus and Nitrosospira* and subgenera *Nitrosobolus* and *Nitrosovibrio* [50, 51].

Studies show that N_2O was emitted from animal houses at the rate of 4–5 mg N m^{-2} d^{-1}, with straw as bedding material, whereas when no bedding material was used, little or no N_2O was emitted from slurry-based cattle or pig building as complete anaerobic condition would have maintained [52]. Deep litter system with fattening pigs showed much higher emission compared to slurry based pig houses, while mechanical mixing still further increased N_2O emission [53]. In cattle collection yards, there had been very less or no N_2O emission as the anaerobic condition prevents conversion of NH_4+ to NO_3-.

Stored solid manures acts as a source of N_2O production/consumption and emission. Covering heaped manure shows reduction in NH_3 emissions but has no effect on N_2O emission, while other studies showed that both were reduced. The addition of chopped straw reduced N_2O emission by 32% from the small scale of cattle manure. [54]. Slurry or liquid manure with no cover showed negligible N_2O release, while slurry with straw cover might act as a source of emission [55]. N_2O emission occurs following manure application to soil [56]. Various factors that affect N_2O release from soil include (i) type of manure, (ii) soil type, (iii) manure composition, (iv) measurement period, (v) timing of manure application, (vi) amount of manure applied, and (vii) method of application.

3.3. Other sources of GHGs from livestock farm

If all parts of the livestock production lifecycle are included, livestock are estimated to account for 18% of global anthropogenic emissions [57]. According to Gill and coworkers [57], apart from enteric fermentation and manure management, the other sources of GHG emission from livestock and related activities are fossil fuels used during feed and fertilizer production and transport of processed animal products.

4. Models for forecasting the greenhouse gas emission in livestock farms

Agricultural production is recognized as a significant contributor to GHG production. Intensive dairy production, in particular, contributes to significant quantities of CH_4 and several forms of nitrogen (N), which can contribute to N_2O production. Over the past 10 years, research studies have attempted to address various sources of GHG emissions within the dairy production system. These sources have included housing [58], manure removal, storage, and treatment systems [59]. Others have compared GHG emissions from conventional farming practices to those employed in organic production. Many of these studies have looked at one section of the production chain in isolation. However, dairy production is a complex system involving inputs such as feed and fertilizer, animals with inherent physiological structures for fermentation of feedstuffs, and the production of manure, storage systems, cropping systems, and export of meat and milk.

It is very easy to understand that attempting to design and conduct research trials to ascertain the effect of one or multiple changes on production, economics, and GHG emissions from a dairy production system would be expensive and time consuming. Therefore, the use of whole farm models, with short-term studies for validation, is an attractive alternative. The integrated farm system model (IFSM) apart from evaluating alternate agronomic feeding, manure storage, and disposal strategies, also accounts for fossil fuel used in farming activities. In real sense, these models do not predict production of GHG but assist in generating some basic information required to predict GHG based on published data.

The development of whole-farm approaches for the mitigation of GHG emissions has been taken up recently by several research groups. A common feature of whole farm models is the ability to calculate CH_4 and N_2O emissions from all farm activities. Furthermore, the models vary considerably on many other aspects. General characteristics of whole farm models include model type, CH_4 and N_2O emissions, CO_2 emissions, C sequestration, NH_3 and NO_3 emissions, P cycling, pre chain emissions, animal welfare, economics, biodiversity, product quality, soil quality, and landscape aesthetics [60]. Whole farm model (WFM) uses pasture growth and cow metabolism for predicting CH_4 emissions in dairy farms. Also included in the WFM is climate and management information. However, recent reports also suggests that WFMs may incorrectly estimate CH_4 emission levels as they do not take into account the DMI and diet composition while predicting the enteric CH_4 emission. This low prediction efficiency of WFMs may lead to substantial error in GHG inventories [10, 11].

The integrated farm system model (IFSM) is a simulation model that integrates the major biological and physical processes of a crop, beef, or dairy farm and evaluating the overall impact of management strategies used to reduce CH_4 emissions [61, 62]. The IFSM is a process-based whole farm simulation including major components for soil processes, crop growth, tillage, planting and harvest operations, feed storage, feeding, herd production, manure storage, and economics [63]. IFSM predicts the effect of management scenarios on farm performance, profitability, and environmental pollutants such as nitrate leaching, ammonia volatilization, and phosphorus runoff loss. The dairy greenhouse gas model (DairyGHG) is a type of IFSM that was developed to provide an easy to use software tool for estimating GHG emissions and the carbon footprint of dairy production systems [64]. Recently, FAO developed a global livestock environmental assessment model (GLEAM), which reported that livestock-related activities contributed around 7.1 gigatonnes CO_2-eq per annum, indicating the prominent contribution of livestock to climate change [65].

A whole-farm approach is a powerful tool for the development of cost-effective GHG mitigation option. The modeling technology can be used to assess the technical, environmental, and financial implications of alternative farm management strategies, under changing external conditions. Whole farm models (WFMs) can reveal relevant interactions between farm components and is useful for integrated scenario development and evaluation. Further, the whole-farm approach ensures that the potential negative trade-offs are taken into account and that positive synergies are identified. In addition, the whole farm models are also used to explore future farm strategies, and since it is operated on farming level, it also provides opportunity for farmers to learn and understand the underlying processes on their own farm. Hence, the whole-farm approach is also helpful in communicating the mitigation option to the farmers, and this could be more beneficial if the models additionally evaluate costs and benefits associating with farming activities.

5. Conclusion

Livestock undoubtedly need to be a priority focus of attention as the global community seeks to address the challenge of climate change. Livestock contribute directly as well as indirectly to global GHG pool. The two primary sources of GHG from livestock are enteric fermentation and manure management. There are several techniques available to quantify CH_4 emission, and the application of appropriate technique depends on objectives of the study. Further, simulation models offer a great scope to predict accurately the GHG emission in farm as a whole. This information will be very valuable in understanding the role of livestock to climate change in depth, and this understanding will help in designing suitable mitigation strategies to reduce livestock-related GHGs.

Author details

Veerasamy Sejian[1,2*], Raghavendra Bhatta[1], Pradeep Kumar Malik[1], Bagath Madiajagan[1], Yaqoub Ali Saif Al-Hosni[2], Megan Sullivan[2] and John B. Gaughan[2]

*Address all correspondence to: drsejian@gmail.com

1 ICAR-National Institute of Animal Nutrition and Physiology, Adugodi, Bangalore, Karnataka, India

2 School of Agriculture and Food Sciences (Animal Science) The University of Queensland, Gatton, QLD, Australia

References

[1] Intergovernmental Panel on Climate Change (IPCC). Climate Change: Synthesis Report; Summary for Policymakers. Retrieved from: http://www.ipcc. ch/pdf/assessment-report/ar4/syr/ar4_syr_spm.pdf. 2007.

[2] Food and Agriculture Organization of the United Nations (FAO). Livestock a Major Threat to the Environment: Remedies Urgently Needed. Retrieved from: http://www.fao.org/newsroom/en/news/2006/ 1000448/index.html. 2006.

[3] Gleik PH, Adams RM, Amasino RM. Climate change and the integrity of science. Science. 2010; 328: 689–691.

[4] Sejian V, Gaughan J, Baumgard L, Prasad CS. Climate Change Impact on Livestock: Adaptation and Mitigation. Springer-Verlag GMbH Publisher, New Delhi, India, 2015a; pp. 1–532.

[5]   Malik PK, Bhatta R, Takahashi J, Kohn RA, Prasad CS. Livestock production and climate change. CABI Climate Change Series 6, CABI Nosworthy Way, Wallingford, UK. 2015a, pp. 1–408.

[6]   Moss AR, Jounany JP, Neevbold J. Methane production by ruminants: its contribution to global warming. Ann. Zootech. 2000; 49: 231–253.

[7]   Naqvi SMK, Sejian V. Global climate change: role of livestock. Asian Journal of Agricultural Sciences. 2011; 3(1): 19–25.

[8]   Bhatta R, Malik PK, Prasad CS. Enteric methane emission: status, mitigation and future challenges—an Indian perspective (chapter 15). In: Malik PK, Bhatta R, Takahashi J, Kohn RA, and Prasad CS (eds.), *Livestock Production and Climate Change*. Publisher CABI, Oxfordshire, UK & CABI, Boston, MA, USA. 2015; pp. 229–244.

[9]   Food and Agriculture Organization of the United Nations (FAO). Submission to UNFCCC AWG LCA, Enabling Agriculture to Contribute to Climate Change. Retrieved from: http://unfccc.int/resource/ docs/2008/smsn/igo/036.pdf. 2009.

[10] Sejian V, Indu S, Ujor V, Ezeji T, Lakritz J, Lal R. Global climate change: enteric methane reduction strategies in livestock. In: Environmental stress and amelioration in livestock production. Sejian V, Naqvi SMK, Ezeji T, Lakritz J, and Lal R (eds.), Springer-Verlag GMbH Publisher, Germany. 2012; pp. 469–502.

[11] Sejian V, Lal R, Lakritz J, Ezeji T. Measurement and Prediction of Enteric Methane Emission. International Journal of Biometeorology. 2011a; 55: 1–16.

[12] Food and Agriculture Organization of the United Nations (FAO). Tackling Climate through Livestock: A Global Assessment of Emissions and Mitigation Opportunities, Rome: FAO. 2013.

[13] Intergovernmental Panel on Climate Change (IPCC). Climate Change 2014: Mitigation of Climate Change, Contribution of Working Group III to the Fifth Assessment Report of the Intergovernmental Panel on Climate Change. Edenhofer O, Pichs-Madruga R, Sokona Y, Farahani E, Kadner S, Seyboth K, Adler A, Baum A, Brunner A, Eickemeier P, Kriemann B, Savolainen J, Schlömer S, von Stechow C, Zwickel T, and Minx JC (eds.), Cambridge University Press, Cambridge. 2014.

[14] Popp A, Lotze-Campen H, Bodirsky B. Food consumption, diet shifts and associated non-CO_2 greenhouse gases from agricultural production. Global Environmental Change. 2010; 20: 451–62.

[15] Garg A, Shukla PR. Emission Inventory of India. Tata McGraw Hill Publishing Co Ltd., New Delhi. 2002. p. 84.

[16] Ahuja D. Climate Change. Technical series U S, UPA Report, 1990.

[17] ALGAS. Asia least-cost greenhouse gas abatement strategy: India, ADB–GEF–UNDP, Asian Development Bank and United Nations Development Programme, Manila, the Philippines, 1998, pp. 238.

[18] Singh GP. Methanogenesis and production of greenhouse gases under animal husbandry system. Report of AP Cess Fund Project, National Dairy Research Institute, Karnal, India. 1998.

[19] Environmental Protection Agency (EPA). Inventory of U S greenhouse gas emissions and sinks: 1990–2000, U S Environmental Protection Agency, EPA 430–R–02–003. 1994.

[20] Swamy M, Bhattacharya, S. Budgeting anthropogenic greenhouse gas emission from Indian livestock using country specific emission coefficients. Current Science. 2006; 91:1340–53.

[21] Jha AK, Singh K, Sharma C, Singh SK, Gupta PK. Assessment of methane and nitrous oxide emissions from livestock in India. Earth Science and Climatic Change. doi.org/10.4172/2157-7617.1000107. 2011.

[22] Chhabra A, Manjunath KR, Panigrahy S, Parihar JS. Spatial pattern of methane emission from Indian livestock. Current Science. 2009; 96: 683–89.

[23] Singh S, Kushwaha BP, Nag SK, Bhattacharya S, Gupta PK Mishra AK, Singh A. Assessment of enteric methane emission of Indian livestock in different agro-ecological regions. Current Science. 2012; 102: 1017–1027.

[24] Patra AK. Trends and projected estimates of GHG emissions from Indian livestock in comparisons with GHG emissions from world and developing countries. Asian-Australasian Journal of Animal Science. 2014. 27: 592–599.

[25] Kamra DN. Enteric methane mitigation: present status and future Prospects. In: Proceedings of Global Animal Nutrition Conference (2014) held at Bangalore, 20–22 April, 2014. pp. 77–87.

[26] Malik PK, Bhatta R, Soren NM, Sejian V, Mech A, Prasad KS, Prasad CS. Feed-based approaches in enteric methane amelioration. In: Malik PK, Bhatta R, Takahashi J, Kohn RA, and Prasad CS (eds.), Livestock production and climate change. CABI Climate Change Series 6, CABI Nosworthy Way, Wallingford, UK. 2015b; pp. 336–359.

[27] Hegarty RS, Gerdes R. Hydrogen production and transfer in the rumen. Recent Advances in Animal Nutrition in Australia, 1999; p. 12.

[28] anssen PH, Kirs M. Structure of the archaeal community of the rumen. Applied Environmental Microbiology. 2008; 74: 3619–3625.

[29] Bhatta R, Enishi O, Kurihara M. Measurement of methane production from ruminants-a review. Asian Australasian Journal of Animal Sciences. 2006a; 20: 1305–1318.

[30] Bhatta R, Tajima K, Kurihara M. Influence of temperature and pH on fermentation pattern and methane production in the rumen simulating fermenter (RUSITEC). Asian Australasian Journal of Animal Sciences. 2006b; 19 (3): 376–380.

[31] Johnson KA, Johnson DE. Methane emissions from cattle. Journal of Animal Science. 1995; 73: 2483–2492.

[32] Johnson KA, Westberg HH, Lamb BK, Kincaid RL. The use of sulphur hexafluoride for measuring methane emissions from farm animals. In Proceedings of the 1st International Conference on Greenhouse Gases and Animal Agriculture, Obihiro, Hokkaido, Japan. 2001. pp. 72–81.

[33] Getachew G, Blummel M, Makkar HPS, Becker K. In vitro gas measuring techniques for assessment of nutritional quality of feeds: a review. Animal Feed Science and Technology. 1998; 72: 261–281.

[34] Wilkins JR. Pressure transducer method for measuring gas production by micro organisms. Applied Microbiology. 1974; 27: 135–140.

[35] Blummel M, Orskov ER. Comparison of gas production and nylon bag degradability of roughages in predicting feed intake in cattle. Animal Feed Science and Technology. 1993; 40:109–119.

[36] Makkar HPS, Blummel M, Becker K. Formation of complexes between polyvinyl pyrrolidones or polyethylene glycols and tannins, and their implication in gas production and true digestibility in *in vitro* techniques. British Journal of Nutrition.1995; 73: 897–913.

[37] Menke KH, Raab L, Salewski A, Steingass H, Fritz D, Schneider W. The estimation of the digestibility and metabolizable energy content of ruminant feeding stuffs from the gas production when they are incubated with rumen liquor. Journal of Agricultural Science. 1979; 93: 217–222.

[38] Beuvink JMW, Spoelstra SF, Hogendorp RJ. An automated method for measuring time-course of gas production of feedstuff incubated with buffered rumen fluid. Netherland Journal of Agricultural Science. 1992; 40: 401–407.

[39] Waghorn GC, Stafford KJ. Gas production and nitrogen digestion by rumen microbes from deer and sheep. New Zealand Journal Agricultural Research. 1993; 36: 493–497.

[40] Theodorou MK, Williams BA, Dhanoa MS, McAllan AB, France J. A simple gas production method using a pressure transducer to determine the fermentation kinetics of ruminant feeds. Animal Feed Science and Technology 1994; 48: 185–197.

[41] Pell AN, Schofield P. Computerised monitoring of gas production to measure forage digestion. Journal of Dairy Science. 1993. 76: 1063–1073.

[42] Cone JW, Gelder AH, Visscher GJW, Oudshoorn L. Influence of rumen fluid and substrate concentration on fermentation kinetics measured with fully automated time related gas production apparatus. Animal Feed Science and Technology. 1996; 61: 113–128.

[43] Schnürer A, Jarvis A. Microbial Handbook for Biogas Plants, Swedish Waste Management U2009: 2010; 03.

[44] Chadwick D, Sommer S, Thorman R, Fangueiro D, Cardenas L, Amon B, Brook TM. Manure management: implications for greenhouse gas emissions. Animal Feed Science and Technology. 2011; 166:514–531.

[45] Bull P, McMillan C, Yamamoto A. Michigan Greenhouse Gas Inventory 1990 and 2002; Report No. CSS05–07. Centre for Sustainable Systems, University of Michigan: City, MI, USA. 2005.

[46] Intergovernmental Panel on Climate Change (IPCC). Greenhouse Gas Inventory Reporting Instructions. Revised 1996 IPCC Guidelines for National Greenhouse Gas Inventories; UNEP, WMO, OECD and IEA: Bracknell, UK. 1997.

[47] Duxbury JM, Harper LA, Mosier AR. Contributions of agroecosystems to global climate change. Agricultural Ecosystem Effects on Trace Gases and Global Climate Change. American Society of Agronomy. 1993. pp. 1–18.

[48] Zaman M, Blennerhassett JD. Effects of the different rates of urease and nitrification inhibitors on gaseous emissions of ammonia and nitrous oxide, nitrate leaching and pasture production from urine patches in an intensive grazed pasture system. Agriculture Ecosystems and Environment. 2010; 136: 236–246.

[49] Zaman M, Nguyen ML. Effect of lime or zeolite on N_2O and N emissions from a pastoral soil treated with urine or nitrate-N fertiliser under field conditions. Agriculture Ecosystems and Environment. 2010; 136: 254–261.

[50] Bremner JM, Blackmer AM. Terrestrial nitrification as a source of atmospheric nitrous oxide. In: Delwiche CC (ed.), Denitrification, Nitrification and Atmospheric Nitrous Oxide. Willey and Sons, New York, 1981. pp. 151–170.

[51] Watson SW, Valos FW, Waterbury JB. The family nitrobacteraceae. In: Starr MP, Stolp H, Trupe HG, Below AP, Shlegel HG (eds.), The Prokaryotes, A handbook on Habits, Isolation, and Identification of Bacteria. Springer-Verlag, Berlin. 1981.

[52] Thorman RE, Harrison R, Cooke SD, Chadwick DR, Burston M, Balsdon SL. Nitrous oxide emissions from slurry- and straw-based systems for cattle and pigs in relation to emissions of ammonia. In: McTaggart I and Gairns L (eds.), Proceedings of SAC/SEPA Conference on Agriculture, Waste and the Environment. Edinburgh (UK), 26–28 March 2002. pp. 26–32.

[53] Groenestein CM, Van Fassen HG. Volatilisation of ammonia, nitrous oxide and nitric oxide in deep-litter systems for fattening pigs. Journal of Agricultural Engineering and Research. 1996; 65: 269–274.

[54] Yamulki S. Effect of straw addition on nitrous oxide and methane emissions from stored farmyard manures. Agriculture Ecosystems and Environment. 2006; 112: 140–145.

[55] Sommer SG, Petersen SO, Sogaard HT. Greenhouse gas emission form live-stock slurry. Journal of Environmental Quality. 2000; 29: 744–751.

[56] Sistani KR, Warren JG, Lovanh N, Higgins S, Shearer S. 2010. Greenhouse gas emissions from swine effluent applied to soil by different methods. Soil Science Society of America Journal. 74: 429–435.

[57] Gill M, Smith P, Wilkinson JM. Mitigating climate change: the role of domestic live-stock. Animal. 2010; 4(3): 323–33.

[58] Ellis JL, Kebreab E, Odongo NE, McBride BW, Okine EK, France J. Prediction of methane production from dairy and beef cattle. Journal of Dairy Science. 2007; 90: 3456–3467.

[59] Amon B, Kryvoruchko V, Amon T, Zechmeister-Boltenstern S. Methane, nitrous oxide and ammonia emissions during storage and after application of dairy cattle slurry and influence of slurry treatment. Agriculture, Ecosystems and Environment. 2006; 112: 153–162.

[60] Sejian V, Hyder I, Ezeji T, Lakritz J, Bhatta R, Ravindra JP, Prasad CS, Lal R. Global warming: role of livestock. In: Sejian V, Gaughan J, Baumgard L, Prasad CS (eds.), Climate Change Impact on Livestock: Adaptation and Mitigation. Springer-Verlag GMbH Publisher, New Delhi, India, 2015b; p 141–170.

[61] Sejian V. Climate change, Green house gas emission and sheep production. In: Shinde AK, Swarnkar CP, and Prince LLL (eds.), Status Papers on Future Research in Sheep Production and Production Development, 50 Years Research Contributions(1962–2012), published by Director, Central Sheep and Wool Research Institute, Avikanagar, Rajasthan-304501, 2012; p. 155–178.

[62] Sejian V, Naqvi SMK. Livestock and climate change: mitigation strategies to reduce methane production. In: Guoxiang Liu (ed.), Greenhouse Gases—Capturing, Utilization and Reduction. Intech Publisher, Croatia, 2012; pp. 254–276.

[63] Rotz CA, Corson MS, Chianese DS, Coiner CU. The integrated farm system model: reference manual. University Park, PA: USDAARS Pasture Systems and Watershed Management research unit: www.ars.usda.gov/SP2UserFiles/Place/19020000/ifsmreference.2009.

[64] Sejian V, Rotz A, Lakritz J, Ezeji T, Lal R. Modeling of green house gas emissions in dairy farms. Journal of Animal Science Advances. 2011b; 1(1): 12–20.

[65] Gerber PJ, Steinfeld H, Henderson B, Mottet A, Opio C, Dijkman J, Falcucci A, Tempio G. Tackling climate change through livestock—a global assessment of emissions and mitigation opportunities. Food and Agriculture Organization of the United Nations (FAO), Rome. 2013.

Effect of Dopants on the Properties of Zirconia-Supported Iron Catalysts for Ethylbenzene Dehydrogenation with Carbon Dioxide

Maria do Carmo Rangel, Sirlene B. Lima,

Sarah Maria Santana Borges and

Ivoneide Santana Sobral

Additional information is available at the end of the chapter

Abstract

Due to the harmful effects of carbon dioxide to the environment, a lot of work has been carried out aiming to find new applications, which can decrease the emissions or to capture and use it. An attractive application for carbon dioxide is the synthesis of chemicals, especially for producing styrene by ethylbenzene dehydrogenation, in which it increases the catalyst activity and selectivity. In order to find efficient catalysts for the reaction, the effect of cerium, chromium, aluminum, and lanthanum on the properties of zirconia-supported iron oxides was studied in this work. The modified supports were prepared by precipitation and impregnated with iron nitrate. The obtained catalysts were characterized by thermogravimetry, Fourier transform infrared spectroscopy, X-ray diffraction, specific surface area measurement, and temperature-programmed reduction. The catalysts showed different textural and catalytic properties, which were associated to the different phases in the solids, such as monoclinic or tetragonal zirconia, hematite, maghemite, cubic ceria, monoclinic or hexagonal lantana, and rhombohedral chromia, the active phases in ethylbenzene dehydrogenation. The most promising dopant was cerium, which produces the most active catalyst at the lowest temperature, probably due to its ability of providing lattice oxygen, which activates carbon dioxide and increases the reaction rate.

Keywords: carbon dioxide, styrene, ethylbenzene, dehydrogenation, zirconia, iron oxide

1. Introduction

3 Although greenhouse gas emissions are reaching alarming rates, 80% of the world's energy consumption still comes from fossil fuels, which have been pointed out as the largest source of carbon dioxide emissions [1]. Over the past decade, the global emissions of carbon dioxide from fossil fuels have increased by 2.7% each year and currently are 60% above the levels registered in 1990, which is considered the reference year for the Kyoto Protocol [2]. On the other hand, it is expected that carbon dioxide emissions should reduce by at least 50% to limit the rise of the global average temperature to 2°C by 2050 [3]. Nowadays, the major environmental concerns worldwide, global warming and the acidification of the oceans, are mainly ascribed to the increase of carbon dioxide concentration [4, 5]. Therefore, several alternatives have been proposed to decrease the carbon dioxide concentration and then to mitigate the environment changes. They include demand-side conservation, supply-side efficiency improvement, increasing reliance on nuclear and renewable energy, and carbon capture and storage (CCS) systems [6]. Among them, CCS is considered the most practical approach for long-term carbon dioxide emission reductions, since fossil fuels will continue to be the major source of energy in the next future. However, there are still some technical and economic barriers to be overcome before it can be used on a large scale. One of the main obstacles is the required large capital investment, besides technical difficulties, such as carbon dioxide leakage rates and limited geological storage capacity. Other drawbacks include the costs of transportation and injection when carbon dioxide is only available offshore, such as in United Kingdom, Norway, Singapore, Brazil, and India [7, 8]. Therefore, a more suitable alternative is to capture and use carbon dioxide (carbon capture and utilization [CCU]), changing the waste carbon monoxide emissions into valuable products such as chemicals and fuels, while contributing to climate change mitigation [9].

Captured carbon dioxide can be used as a commercial product, both directly or after conversion. In food and drink industries, for instance, carbon dioxide is often used as a carbonating agent, preservative, packaging gas, and for extracting flavors, as well as in the decaffeination process. In the pharmaceutical industry, it is used as a respiratory stimulant or for the synthesis of drugs. However, these applications are restricted to high-purity carbon dioxide, as that obtained in ammonia plants [9, 10]. Moreover, pressurized carbon dioxide has been investigated for wastewater treatment and water disinfection [11]. Other direct applications of carbon dioxide include enhanced oil recovery and coal-bed methane recovery, where crude oil is extracted from an oil field or natural gas from unminable coal deposits [9].

In the production of chemicals and fuels, carbon dioxide has attracted increasing attention over several decades, for the synthesis of various fine and bulk chemicals. It has already been used in the industrial production of urea, cyclic carbonates, salicylic acid, and methanol [12]. It is expected that carbon dioxide can produce feedstock for chemical, pharmaceutical, and polymer industries by carboxylation reactions to obtain organic compounds, such as carbonates, acrylates, and polymers, or by reduction reactions, where the C=O bonds are broken to produce chemicals such as methane, methanol, syngas, urea, and formic acid [9, 13]. Carbon dioxide can have several other applications, both as carbon or oxygen sources, for the synthesis of chemicals by several processes, as solvent and/or as reactants. It has potential applications in supercritical conditions, in direct carboxylation reactions, in the conversion of natural gas to liquid (GTL technology), and in methanol synthesis [14]. Carbon dioxide can also act as an oxidant in the dehydrogenation of ethane [15], propane [16], isobutene [17], and ethylbenzene [18–20], as well as in methane dry reforming [21] and oxidative coupling of methane [22]. It is expected that the 115 million metric tons of carbon dioxide, currently consumed every year as feedstock in a variety of synthetic processes, can be triplicated by the use of new technologies [19]. In addition, carbon dioxide can overcome several drawbacks of the processes, especially in the case of dehydrogenation reactions.

In industrial processes, the dehydrogenation of hydrocarbons is often carried out at high temperatures to increase the conversion because of its reversibility and limitation by thermo-dynamic equilibrium. Besides being an energy-consuming process, the high temperatures cause the hydrocarbons cracking, decreasing the selectivity. On the other hand, by the oxidation of the produced hydrogen or by using an oxidant in the presence of a catalyst, these difficulties can be overcome, since the oxidative dehydrogenation is exothermic and can be performed at low temperatures, making negligible the formation of cracking products. Therefore, the use of an oxidant increases the catalyst selectivity and decreases the undesirable products, besides other advantages. Among the oxidizing agents, carbon dioxide has proven to be the most promising one for dehydrogenation reactions [23]. In the ethylbenzene dehy-drogenation, for instance, the use of carbon dioxide can provide a route, which represents an elegant and promising alternative to the conventional process of styrene production.

Currently, the ethylbenzene dehydrogenation in the presence of overheated steam [Eq. (1)] is the main commercial route to produce styrene, one of the most used intermediate for organic synthesis. It is the main building block for several polymers, such as polystyrene, styrene-butadiene rubber, styrene-acrylonitrile, acrylonitrile-butadiene-styrene, and other high-value products. The ethylbenzene dehydrogenation supplies 90% of the global production of styrene, which was around 30×10^6 t in 2010 [24].

$$C_6H_5CH_2CH_{3(g)} + H_2O_{(g)} \longrightarrow C_6H_5CH = CH_{2(g)} + H_{2(g)} \tag{1}$$

In spite of this fact, the commercial process still has several drawbacks, such as the high consumption of energy, the reaction endothermicity ($\Delta H = 124.85$ kJ/mol), the equilibrium limitation of reaction, and the catalyst deactivation [25]. On the other hand, the replacement of steam by carbon dioxide leads to a consumption of $1.5–1.9 \times 10^8$ cal, instead of 1.5×10^9 cal/mol of styrene produced. In this case, hydrogen is continuously removed as steam by the reverse water gas shift reaction, and the equilibrium is shifted to the formation of dehydro-genation products [Eq. (2)]. In addition, carbon dioxide removes the coke deposits formed during the reaction [26].

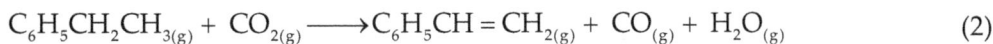

$$C_6H_5CH_2CH_{3(g)} + CO_{2(g)} \longrightarrow C_6H_5CH = CH_{2(g)} + CO_{(g)} + H_2O_{(g)} \tag{2}$$

The use of carbon dioxide includes other advantages such as being an inexpensive, nontoxic, and renewable feedstock, which provides a positive impact on the global carbon balance. In addition, it can accelerate the reaction rate, improve styrene selectivity, decrease the thermo-dynamic limitations, suppress the total oxidation, increase the catalyst life, and avoid hotspots [27]. Therefore, the ethylbenzene dehydrogenation with carbon dioxide has been studied over several different catalysts, including iron oxide, vanadium oxide, antimony oxide, chromium oxide, cerium oxide, zirconium oxide, lanthanum oxide, perovskites, and the oxide catalysts promoted with alkali metals supported on several oxides [16, 19, 20, 24, 26–33]. In addition, several works have shown that carbon-based catalysts are active and selective to produce styrene through ethylbenzene dehydrogenation with carbon dioxide. Activated carbons [34, 35], carbon nanofibers [36], onion-like carbons [37], diamonds and nanodiamonds [37, 38], graphites [39], and multiwalled carbon nanotubes (MWCNTs) [40], among others, have been evaluated in ethylbenzene dehydrogenation.

These studies have shown that the effect of carbon dioxide on the activity, selectivity, and stability of the catalysts for ethylbenzene dehydrogenation depends on the kind of the catalyst, as well as on the reaction conditions. For zirconia-based catalysts, the positive effect of carbon dioxide was found to be highly dependent on the crystalline phase at 550°C. It was noted that the tetragonal phase showed high activity and selectivity to styrene, a fact that was related to differences in specific surface area of the solids and their affinity with carbon dioxide associated with the surface basic sites [41, 42]. In a previous work [19], we have found that zirconia was the most active and selective catalyst to produce styrene through ethylbenzene dehydrogen-ation with carbon dioxide, as compared to metal oxides such as lantana (La_2O_3), magnesia (MgO), niobia (Nb_2O_5), and titania (TiO_2). This finding was related to the highest intrinsic activity of zirconia.

In spite of the numerous studies on the catalyst properties for the dehydrogenation of ethylbenzene in the presence of carbon dioxide, no satisfactory catalyst was found yet, requiring new developments. In the present work, the effect of cerium, chromium, aluminum, and lanthanum on the properties of zirconia-supported iron oxides was studied aiming to find efficient catalysts for the reaction.

2. Experimental

2.1. Catalysts preparation

The precursor of zirconium oxide was obtained by hydrolysis of zirconium oxychloride (1 mol/l) with an ammonium hydroxide solution (30% w/v). The obtained gel was rinsed with an ammonium hydroxide solution (1% w/v) eight times up not to detect chloride ions by Mohr's method anymore. The gel was then dried in an oven at 120°C, for 12 h. The solid was calcined at 600°C, for 4 h, under airflow (50 ml/min).

The metal-doped zirconia samples were prepared by the same method, using solutions of zirconium oxychloride and of metal nitrates (Zr/M = 10), where M = Ce (FCEZ sample), Cr

(FCRZ sample), Al (FALZ sample and La (FLAZ sample). Cerium, chromium, aluminum, and lanthanum oxides were also prepared following the same procedure, using aluminum nitrate, cerium nitrate, lanthanum nitrate, and chromium nitrate, respectively, to be used as references.

The modified zirconium oxides were subsequently impregnated with an iron nitrate solution (0.17 mol/l), at room temperature, to obtain the catalysts.

2.2. Catalysts characterization

After iron impregnation, the samples (catalyst precursors) were analyzed by thermogravimetry (TG) and Fourier transform infrared spectroscopy (FTIR).

After calcination, the catalysts were characterized by Fourier transform infrared spectroscopy, X-ray diffraction (XRD), specific surface area measurement, and temperature-programmed reduction.

The experiments of thermogravimetry (TG) were performed on a Mettler Toledo TGA/SDTA 851 equipment. The sample (0.02 g) was placed in a platinum crucible and heated (10°C/min) from room temperature to 1000°C, under airflow (50 ml/min).

The presence of nitrate species in the samples was detected by FTIR, using a Perkin Elmer, Model—Spectrum One, equipment, in the range of 400–4000 cm^{-1}. The samples were prepared as potassium bromide discs, in a 1:10 proportion.

The experiments of X-ray diffraction (XRD) were carried out in a Shimadzu model XD3A apparatus, using CuKα radiation generated at 30 kV and 20 mA and nickel filter.

The specific surface areas were measured in a Micromeritics ASAP 2020, using the sample (0.2 g) previously heated at 300°C, under nitrogen flow.

The curves of temperature-programmed reduction were obtained on a Micromeritics model TPR/TPD 2900 equipment, utilizing 0.3 g of the sample, and heating the solid with a rate of 10°C/min, under flow of a mixture of 5% hydrogen in nitrogen up to 1000°C.

2.3. Catalysts evaluation in ethylbenzene dehydrogenation with carbon dioxide

The catalysts were evaluated in ethylbenzene dehydrogenation in the presence of carbon dioxide in a fixed bed reactor, using 0.3 g of catalyst, at several temperatures (530, 550, 570, 590, 610, and 630°C) under atmospheric pressure. A carbon dioxide to ethylbenzene molar ratio of 10 was used for all experiments.

The reaction products were analyzed by online gas chromatography, using a Varian Star 3600 Cx equipment with a flame ionization detector. A commercial catalyst for the ethylbenzene dehydrogenation with steam, based on iron and chromium oxides, was also evaluated in the same conditions, for comparison.

3. Results and discussion

3.1. Thermogravimetry

The TG curves for the catalyst precursors (before calcination) are displayed in **Figure 1**. For all cases, there was a weight loss in two stages: the first at around 200°C, related to loss of volatiles adsorbed on the solids; the second stage at higher temperatures, in the range of 200–450°C, can be assigned to the decomposition of iron hydroxide to produce hematite and/or maghemite [43, 44]. It can be noted that the kind of the support affected hematite formation, probably due to different interactions of the iron oxide precursor with the support. The process was easier over lanthanum-doped zirconia (225°C), followed by cerium-doped zirconia (250°C). On the other hand, for aluminum-doped zirconia (292°C) and for chromium-doped zirconia (300°C), the process was delayed, suggesting that iron hydroxide was more strongly bonded to these supports.

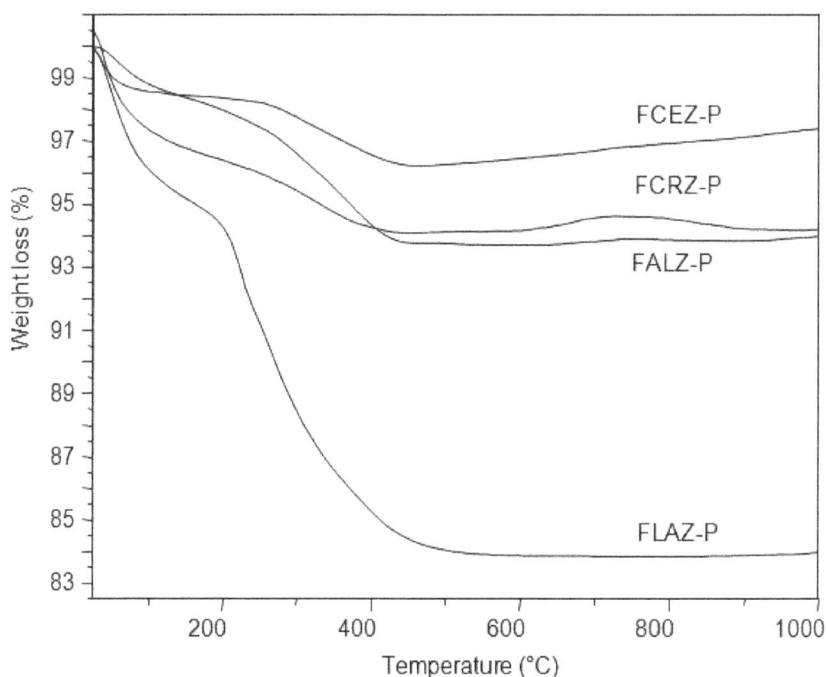

Figure 1. TG curves for the catalyst precursors. F, iron; CE, cerium; CR, chromium; AL, aluminum; LA, lanthanum; Z, zirconia.

3.2. Fourier transform infrared spectroscopy

The FTIR spectra for the precursors (**Figure 2a**) show two bands at 3400 and 1600 cm^{-1}, assigned to the bending vibrations of OH groups in iron hydroxides and in adsorbed water [45]. The absorption at 1384 cm^{-1} is related to the nitrate species [46], from iron nitrate. In the low-frequency region, a broad band was observed, in the range of 800–400 cm^{-1}, attributed to the

Fe–O bond [45]. For the catalysts (**Figure 2b**), it can noted that the band at 1384 cm^{-1} decreased for the samples, except for chromium-doped catalyst, indicating that the calcination was effective for the removal of nitrate species.

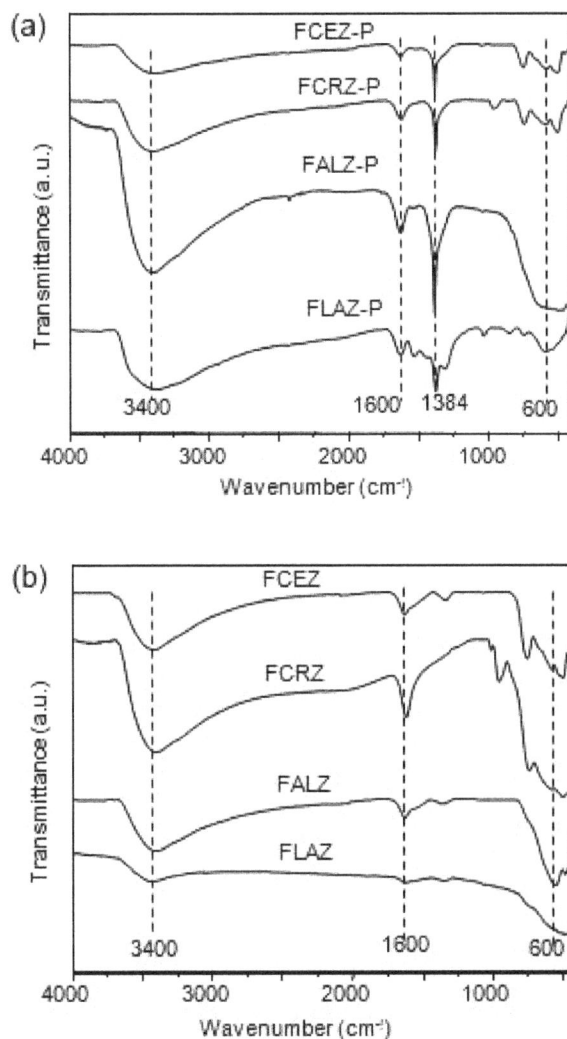

Figure 2. FTIR spectra for the precursors (P) and for the catalysts. F, iron; CE, cerium; CR, chromium; AL, aluminum; LA, lanthanum; Z, zirconia.

3.3. X-ray diffraction

From the X-ray diffractograms of the solids (**Figure 3**), different phases were found for all samples, related to the different oxides. However, for most cases, it was not possible to assure the presence of isolated phases of iron, zirconium, and of the dopants. Therefore, hematite, α-Fe_2O_3 (JCPDS 871166), maghemite, γ-Fe_2O_3 (JCPDS 251402), or zirconium oxide, ZrO_2 (monoclinic, JCPDS 830944 and tetragonal, JCPDS 881007), as well as lanthanum oxide, La_2O_3

(monoclinic, JCPDS 220641 or hexagonal, JCPDS 401279), aluminum oxide, Al_2O_3 (orthorhombic, JCPDS 880107), or chromium oxide, Cr_2O_3 (rhombohedral, JCPDS 841616), cannot be detected, because of the coincidence of the diffraction peaks of these phases. Only maghemite and the cubic phase of ceria, CeO_2 (JCPDS 780694), were detected as isolated phases for the chromium and cerium-doped samples, respectively.

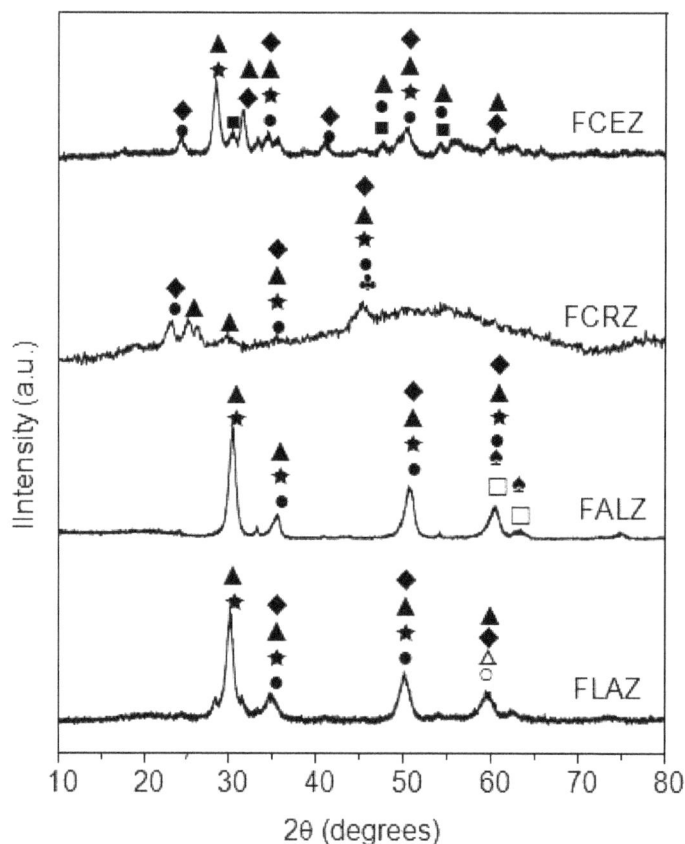

Figure 3. X-ray diffractograms for the catalysts. F, iron; CE, cerium; CR, chromium; AL, aluminum; LA, lanthanum; Z, zirconia. ♦: hematite (α-Fe_2O_3), ematite (ray dγ-Fe_2O_3, ★: tetragonal zirconia (ZrO_2), •, monoclinic zirconia (ZrO_2), ■, cubic ceria (CeO_2), ♣: rhombohedral chromia (Cr_2O_3), △:hexagonal lantana (La_2O_3), ○, monoclinic lantana (La_2O_3), □: orthorhombic alumina (Al_2O_3), ♠: monoclinic alumina (Al_2O_3).

3.4. Specific surface areas

Table 1 shows the specific surface areas of the catalysts, as well as of pure and doped supports. It can be noted that pure oxides showed different values, which are typical of the nature of each oxide. Zirconia showed the highest values, while chromium showed the lowest one. In addition, the dopants changed the specific surface area of zirconia (73 m^2/g), depending on the kind of dopant. These different behaviors are related to the size of the ions, the possibility of the ion to enter into zirconia lattice, and the formation of mixed compounds. According to previous studies [47–49], it would be expected that these dopants would increase the specific surface areas of zirconia, because of the differences in ionic radius of cerium (0.97 Å), chromium (0.615 Å), aluminum (0.54 Å), and lanthanum (1.16 Å), as compared to zirconium (0.84 Å). These differences often cause stresses in zirconia lattice, favoring the production of smaller particles, since they decrease the stress to surface ratio. However, only for the aluminum-doped zirconia the specific surface area increased, suggesting that most of the dopants did not enter into the lattice but rather remain as a segregated phase, as detected for cerium-doped zirconia.

The impregnation of iron on the supports also changes the specific surface areas, as shown in **Table 1**. For the chromium-doped and lanthanum-doped samples, the addition of iron caused an increase in specific surface area, suggesting a contribution of the iron oxides to these values. On the other hand, the other samples showed a decrease in the specific surface area, indicating that they went on sintering during the calcination step, after iron impregnation. The chromium-based catalyst showed the highest value, while the cerium-based catalyst showed the lowest one.

Samples	Sg (m²/g)
Z	73
CE	38
CR	1.7
AL	23
LA	17
CEZ	74
CRZ	17
ALZ	98
LAZ	19
FCEZ	58
FCRZ	127
FALZ	85
FLAZ	95

Z, zirconia; CE, cerium or ceria; CR, chromium or chromia; AL, aluminum or alumina; LA, lanthanum or lanthana; F, iron oxide.

Table 1. Specific surface areas (Sg) of pure oxides, doped zirconia, and of iron oxide supported on doped zirconia.

3.5. Temperature-programmed reduction

The catalysts showed different reduction profiles, as displayed in **Figure 4**. The cerium-doped catalyst showed a peak beginning at 192°C and others in the range from 398 to 931°C. The first peak can be assigned to the reduction of Fe^{+3} to Fe^{+2} species, while the latter is due to the reduction of Fe^{+2} to Fe^0 species [50], as well as to the reduction processes related to the support

[32, 33]. On the other hand, the chromium-doped zirconia sample showed a reduction peak beginning at 182°C, with a shoulder at around 274°C, as well as another peak in the range 501–929°C. The first peak can be associated to the reduction of Cr^{+6} to Cr^{+3} [16] species and of Fe^{+3} to Fe^{+2} species, while the latter one is due to the reduction of Fe^{+2} to species Fe^0 [50]. The lanthanum-doped sample showed a peak beginning at 225°C and other ones at 327, 393, and 500°C, attributed to the reduction of Fe^{+3} species in different interactions with the support. A broad peak in the range of 600–781°C is related to the reduction of Fe^{+2} to Fe^0 species and to the processes related to the support. For the aluminum-doped sample, two reduction peaks beginning at 200 and 332°C were noted associated to the reduction of Fe^{+3} to Fe^{+2} species in different interactions with the support. A broad peak in the range of 406–704°C can be assigned to the reduction of Fe^{+2} to Fe^0 species. The easiness of the reduction decreased with the dopants in the order: Cr > Ce > Al > La.

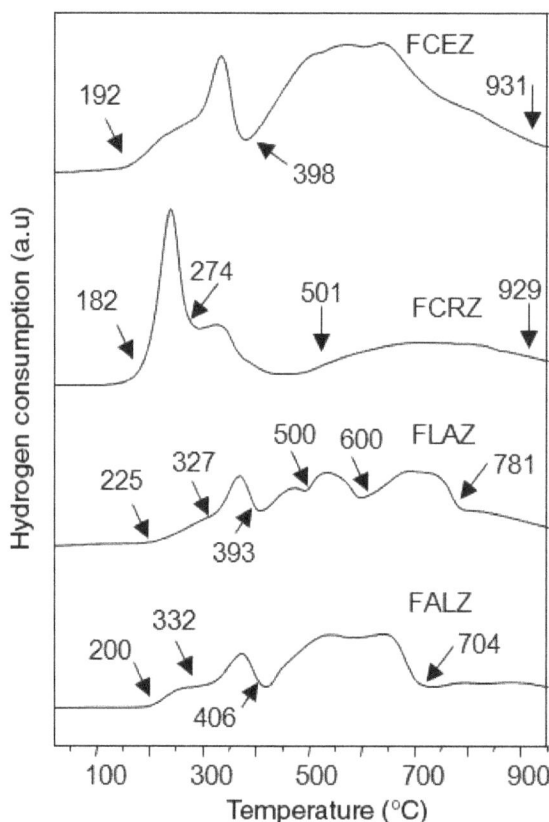

Figure 4. Curves of temperature-programmed reduction for the catalysts. F, iron; CE, cerium; CR, chromium; AL, aluminum; LA, lanthanum; Z, zirconia.

3.6. Activity and selectivity of the catalysts

Figure 5 shows the values of ethylbenzene conversions as a function of temperature during the dehydrogenation with carbon dioxide. It can be noted that the samples were more active than a commercial catalyst, for all temperature ranges. Also, the catalysts showed different performances, depending on the reaction temperature. At low temperatures, the cerium-based sample led to the highest conversion that, however, decreased with the temperature increase. This can be related to the ability of cerium oxide (detected by X-ray diffraction) for providing lattice oxygen, which activates the carbon dioxide molecule and then increases the reaction rate [32, 33]. The chromium-doped catalyst was the second most active one, leading to conversions of around 46%, which increased with temperature, a fact that can be associated to the high dehydrogenation activity of chromium compounds [16]. The aluminum-doped and lanthanum-doped samples showed similar behaviors, leading to low conversions that increased with temperature.

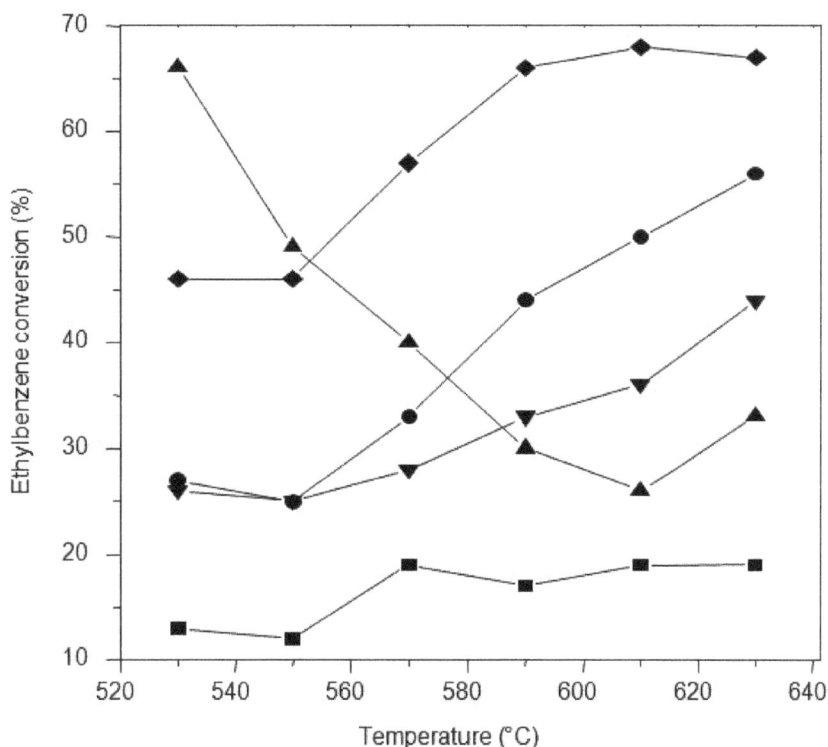

Figure 5. Ethylbenzene conversion over the obtained catalysts and over a commercial catalyst. F, iron; CE, cerium; CR, chromium; AL, aluminum; LA, lanthanum; Z, zirconia.

The selectivity of the catalysts to styrene (**Figure 6**) also changed with the kind of the dopant and with temperature. The aluminum-doped catalyst was the most selective one, but the selectivity decreased as the temperature increased. A similar behavior was noted for the commercial catalyst. On the other hand, the selectivity of cerium-doped sample showed a maximum at around 570°C, while the selectivity of lanthanum-based and chromium-based solids almost did not change with temperature. These findings can be related to the kind of the dopants and their different interactions with the support, as well as to the reaction temperature.

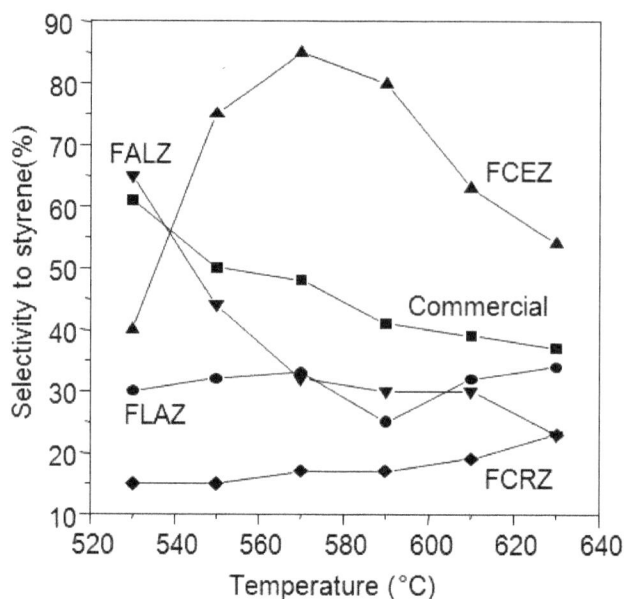

Figure 6. Selectivity to styrene of the obtained catalysts and of a commercial catalyst, during ethylbenzene conversion. F, iron; CE, cerium; CR, chromium; AL, aluminum; LA, lanthanum; Z, zirconia.

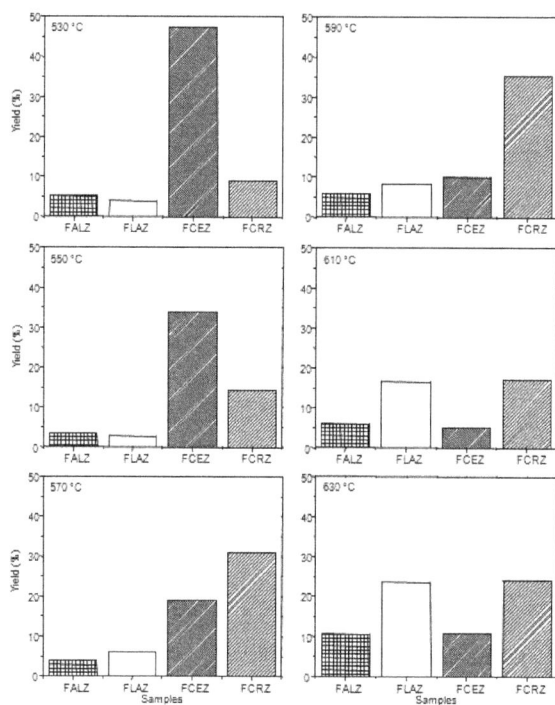

Figure 7. Styrene yields over the obtained catalysts and over a commercial catalyst, during ethylbenzene conversion. F, iron; CE, cerium; CR, chromium; AL, aluminum; LA, lanthanum; Z, zirconia.

Figure 7 shows the yields obtained over the catalysts. One can see that the yield largely depends on the reaction temperature and on the kind of the dopant. The highest value was obtained at 530 and 560°C, over the cerium-doped catalyst. However, the yield decreased with temperature increase, suggesting the catalyst deactivation at high temperatures. While the other catalysts showed low yields for all temperature ranges, the chromium-doped catalyst led to a yield of around 35% at 590°C.

4. Conclusions

Catalysts based on iron oxides (hematite and/or maghemite), supported on zirconium oxide doped with cerium, chromium, aluminum, or lanthanum, show different textural and catalytic properties in ethylbenzene dehydrogenation with carbon dioxide. These findings can be related to the different phases of the supports, such as zirconia (monoclinic or tetragonal), iron oxides (hematite or maghemite), cerium oxide (cubic), lantana (monoclinic or hexagonal), and chromium oxide (rhombohedral), which are also active in the reaction.

The most promising sample was the cerium-doped solid, which led the highest yield (46%) at the lowest temperature. This was assigned to the role of cerium oxide in providing lattice oxygen, which activates carbon dioxide and increases the reaction rate.

The catalysts have proven to provide another alternative to use carbon dioxide, one of the main greenhouse gas and then to contribute to the environment protection.

Acknowledgements

SBL and SMSB acknowledge CAPES and CNPq for their fellowships. The authors thank CNPq and FINEP for the financial support.

Author details

Maria do Carmo Rangel[1,2*], Sirlene B. Lima[1,2], Sarah Maria Santana Borges[1] and Ivoneide Santana Sobral[1]

*Address all correspondence to: mcarmov@ufba.br

1 Grupo de Estudos em Cinética e Catálise, Instituto de Química, Universidade Federal da Bahia, Campus Universitário de Ondina, Salvador, Bahia, Brazil

2 Programa de Pós-Graduação em Engenharia Química, Rua Aristides Novis, Salvador, Bahia, Brazil

References

[1] Abass AO. Valorization of greenhouse carbon dioxide emissions into value-added products by catalytic processes. Journal of CO_2 Utilization. 2013;3-4:74–92. DOI: 10.1016/j.jcou.2013.10.004

[2] Quéré CL, Peters GP, Andres RJ, Andrew RM, Boden T, Ciais P, Friedlingstein P, et al. Global carbon budget. Earth System Science Data Discussion. 2013;6:689–760. DOI: 10.5194/essdd-6-689-2013

[3] IPCC, Climate Change 2013: The Physical Science Basis, Intergovernmental Panelon Climate Change [Internet]. 2013. Available at: https://www.ipcc.ch/report/ar5/wg1/ [Accessed: 2016-02-11].

[4] Honisch B, Ridgwell A, Schmidt DN, Thomas E, Gibbs SJ, Sluijs A, Zeebe R, Kump L, Martindale RC, et al. The geological record of ocean acidification. Science. 2012;335:1058–1063. DOI: 10.1126/science.1208277

[5] Crowley TJ, Berner RA. Enhanced: CO2 and climate change. Science. 2001;292:870–872. DOI: 10.1126/science.1061664

[6] Spigarelli BP, Kawatra SK. Opportunities and challenges in carbon dioxide capture. Journal of CO2 Utilization. 2013;1:69–87. DOI: 10.1016/j.jcou.2013.03.002

[7] Ciferno JP, Fout TE, Jones AP, Murphy JT. Capturing carbon from existing coal fired power plants. Chemical Engineering Progress. 2009;105:33–41. ISSN: 03607275

[8] D'Alessandro DM, Smit B, Long JR. Carbon dioxide capture: Prospects for new materials. Angewandte Chemie International Edition. 2010;49:6058–6082. DOI: 10.1002/anie.201000431

[9] Cuéllar-Franca RM, Azapagic A. Carbon capture, storage and utilisation technologies: A critical analysis and comparison of their life cycle environmental impacts. Journal of CO2 Utilization. 2015;9:82–102, DOI: 10.1016/j.jcou.2014.12.001

[10] Markewitz P, Kuckshinrichs W, Leitner W, Linssen J, Zapp P, Bongartz R, Schreiber A, Müller TE. Worldwide innovations in the development of carbon capture technologies and the utilization of CO_2. Energy and Environmental Science. 2012;5:7281–7305. DOI: 10.1039/C2EE03403D

[11] Vo HT, Imai T, Ho TT, Dang T-L, Hoang SA. Potential application of high pressure carbon dioxide in treated wastewater and water disinfection: Recent overview and further trends. Journal of Environmental Sciences. 2015;36:38–47. DOI: 10.1016/j.jes.2015.04.006

[12] Plasseraud L. Carbon Dioxide as Chemical Feedstock. In: Aresta, M, editor. Chem Sus Chem; 2010. p. 631–632. DOI: 10.1002/cssc.201000097

[13] Boxun H, Curtis G, Steven LS. Thermal, electrochemical, and photochemical conversion of CO_2 to fuels and value-added products. Journal of CO2 Utilization. 2013;1:18–27. DOI: 10.1016/j.jcou.2013.03.004

[14] Aresta M, Dibenedetto A. Product review. The contribution of the utilization option to reducing the CO_2 atmospheric loading: Research needed to overcome existing barriers for a full exploitation of the potential of the CO_2 use. Catalysis Today. 2004;98:455–462. DOI: 10.1016/j.cattod.2004.09.001

[15] Nakagawa K, Okamura M, Ikenaga N, Suzuki T, Kobayashi T. Dehydrogenation of ethane over gallium oxide in the presence of carbon dioxide. Chemical Communications. 1998;9:1025–1026. DOI: 10.1039/A800184G

[16] Wang S, Murata K, Hayakawa T, Hamakawa S, Suzuki K. Dehydrogenation of ethane with carbon dioxide over supported chromium oxide catalysts. Applied Catalysis A. 2000;196:1–8. ISSN: 0926-860X. DOI:10.1016/S0926-860X(99)00450-0

[17] Shimada H, Akazawa T, Ikenaga N, Suzuki T. Dehydrogenation of isobutane to isobutene with iron-loaded activated carbon catalyst. Applied Catalysis A. 1998;168:243–250. ISSN: 0926-860X. DOI:10.1016/S0926-860X(97)00350-5

[18] Mimura N, Takahara I, Saitoa M, Hattorib T, Ohkumac K, Andod M. Dehydrogenation of ethylbenzene over iron oxide-based catalyst in the presence of carbon dioxide. Catalysis Today. 1998;45:61–64. DOI:10.1016/S0920-5861(98)00246-6

[19] Rangel MC, Monteiro AM, Oportus M, Reyes P, Ramos MS, Lima SB. Ethylbenzene dehydrogenation in the presence of carbon dioxide over metal oxides. In: Guoxiang, L, editor. Greenhouse Gases: Capturing, Utilization and Reduction. Publishing Intech; 2012. p. 117–136. ISBN: 978953-51-0192-5

[20] Rangel MC, Monteiro APM, Marchetti SG, Lima SB, Ramos MS. Ethylbenzene dehydrogenation in the presence of carbon dioxide over magnesia-supported iron oxides. Journal of Molecular Catalysis A, Chemical. 2014;387:147–155. DOI:10.1016/j.molcata.2014.03.002

[21] Rangel MC, de Araújo GC, de Lima SM, Assaf JM, Peña MA, Fierro JLG. Catalytic evaluation of perovskite-type oxide $LaNi_{1-x}Ru_xO_3$ in methane dry reforming. Catalysis Today. 2008;133-135:129–135. DOI:10.1016/j.cattod.2007.12.049

[22] Nishiyama T, Aika K. Mechanism of the oxidative coupling of methane using CO_2 as an oxidant over PbO-MgO. Journal of Catalysis. 1990;122:346–351. DOI:10.1016/0021-9517(90)90288-U

[23] Corberán VC. Novel approaches for the improvement of selectivity in the oxidative activation of light alkanes. Catalysis Today. 2005;99:33–41. DOI:10.1016/j.cattod.2004.09.055

[24] Betiha MA, Rabie AM, Elfadly AM, Yehia FZ. Microwave assisted synthesis of a VOx-modified disordered mesoporous silica for ethylbenzene dehydrogenation in pres-

ence of CO_2. Microporous and Mesoporous Materials. 2016;222:44–54. DOI: 10.1016/j.micromeso.2015.10.009

[25] Nederlof C, Kapteijn F, Makkee M. Catalysed ethylbenzene dehydrogenation in CO_2 or N_2-Carbon deposits as the active phase. Applied Catalysis A. 2012;417-418:163–173. DOI:10.1016/j.apcata.2011.12.037

[26] Mamedov EA, Corberan VC. Oxidative dehydrogenation of lower alkanes on vanadium oxide-based catalysts. The present state of the art and outlooks. Applied Catalysis A. 1995;127:1–40. DOI:10.1016/0926-860X(95)00056-9

[27] Sakurai Y, Suzaki T, Nakagawa K, Ikenagaa N-O, Aota H, Suzuki T. Dehydrogenation of ethylbenzene over vanadium oxide-loaded mgo catalyst: Promoting effect of carbon dioxide. Journal of Catalysis. 2002;209:16–24. Doi:10.1006/jcat.2002.3593

[28] Hong D, Vislovskiy VP, Hwang YK, Hung SHJ, Chang J. Dehydrogenation of ethylbenzene with carbon dioxide over MgO-modified Al2O3-supported V-Sb oxide catalysts. Catalysis Today. 2008;131:140–145. DOI:10.1016/j.cattod.2007.10.020

[29] Burri A, Jiang N, Yahyaoui K, Park S-E. Ethylbenzene to styrene over alkali doped TiO2-ZrO_2 with CO_2 as soft oxidant. Applied Catalysis A: General. 2015;495:192–199. DOI: 10.1016/j.apcata.2015.02.003

[30] Zhang S-J, Li W-Y, Li X-H. Effect of preparation methods on the catalytic properties of Fe_2O_3/Al_2O_3-ZrO_2 for ethylbenzene dehydrogenation. Journal of Fuel Chemistry and Technology. 2015;43:437–441. DOI:10.1016/S1872-5813(15)30012-8

[31] Watanabe R, Saito Y, Fukuhara C. Enhancement of ethylbenzene dehydrogenation of perovskite-type $BaZrO_3$ catalyst by a small amount of Fe substitution in the B-site. Journal of Molecular Catalysis A: Chemical. 2015;404-405:57–64. DOI: 10.1016/j.molcata.2015.04.010

[32] Kovacevic M, Agarwal S, Mojet BL, Ommen JGV, Lefferts L. The effects of morphology of cerium oxide catalysts for dehydrogenation of ethylbenzene to styrene. Applied Catalysis A: General. 2015;505:354–364. DOI: 10.1016/j.apcata.2015.07.025

[33] Li X, Feng J, Fan H, Wang Q, Li W. The dehydrogenation of ethylbenzene with CO_2 over $Ce_xZr_{1-x}O_2$ solid solutions. Catalysis Communications. 2015;59:104–107. DOI: 10.1016/j.catcom.2014.10.003

[34] Zarubina V, Talebi H, Nederlof C, Kapteijn F, Makkee M, Cabrera IM. On the stability of conventionaland nano-structured carbon-based catalystsin theoxidative dehydrogenation of ethylbenzeneunder industrially relevant conditions. Carbon. 2014;77:329–340. DOI: 10.1016/j.carbon.2014.05.036

[35] Oliveira SB, Barbosa DP, Monteiro APM, Rabelo D, Rangel MC. Evaluation of copper supported on polymeric spherical activated carbon in the ethylbenzene dehydrogenation. Catalysis Today. 2008;133-135:92–98. ISSN: 09205861. DOI:10.1016/j.cattod.2007.12.040

[36] Delgado JJ, Chen XW, Frank B, Su DS, Schlögl R. Activation processes of highly ordered carbon nanofibers in the oxidative dehydrogenation of ethylbenzene. Catalysis Today. 2012;186:93–98. DOI:10.1016/j.cattod.2011.10.023

[37] Su D, Maksimova NI, Mestl G, Kuznetsov VL, Keller V, Schlögl R, et al. Oxidative dehydrogenation of ethylbenzene to styrene over ultra-dispersed diamond and onion-like carbon. Carbon. 2007;45:2145–2151. DOI: 10.1016/j.carbon.2007.07.005

[38] Ba H, Liu Y, Mu X, Doh W-H, Nhut J-M, Granger P, Pham-Huu C. Macroscopic nanodiamonds/β-SiC composite as metal-free catalysts for steam-free dehydrogenation of ethylbenzene to styrene. Applied Catalysis A: General. 2015;499:217–226. DOI: 10.1016/j.apcata.2015.04.022

[39] Li P, Li T, Zhou JH, Sui ZJ, Dai YC, Yuan WK, et al. Synthesis of carbon nanofiber/graphite-felt composite as a catalyst. Microporous Mesoporous Materials. 2006;95:1–7. DOI:10.1016/j.micromeso.2006.04.014

[40] Qui N, Scholz P, Keller T, Pollok K, Ondruschka B. Ozonated multiwalled carbon nanotubes as highly active and selective catalyst in the oxidative dehydrogenation of ethyl benzene to styrene. Chemical Engineering & Technology. 2013;36:300–306. DOI: 10.1002/ceat.201200354

[41] Vislovskiy VP, Chang J-S, Park M-S, Park S-E. Ethylbenzene into styrene with carbon dioxide over modified vanadia–alumina catalysts. Catalysis Communications. 2002;3:227–231. ISSN: 1566-7367. DOI:10.1016/S1566-7367(02)00105-X

[42] Sun A, Qin Z, Wang J. Reaction coupling of ethylbenzene dehydrogenation with water-gas shift. Applied Catalysis A. 2002;234:179–189. ISSN: 0926-860X. DOI:10.1016/S0926-860X(02)00222-3

[43] Gadalla AM, Livingston TW. Thermal behavior of oxides and hydroxides of iron and nickel. Thermochimica Acta. 1989;145:l–9. DOI:10.1016/0040-6031(89)85121-4

[44] Silva CLS, Marchetti SG, Faro ACJ, Silva TF, Assaf JM, Rangel MC. Effect of gadolinium on the catalytic properties of iron oxides for WGSR. Catalysis Today. 2013;213:127–134. DOI:10.1016/j.cattod.2013.02.025

[45] Pedrosa J, Costa BFO, Portugal A, Durães L. Controlled phase formation of nanocrystalline iron oxides/hydroxides in solution: An insight on the phase transformation mechanisms. Materials Chemistry and Physics. 2015;163:88–98. DOI:10.1016/j.matchemphys.2015.07.018

[46] Martino AD, Iorio M, Prenzler PD, Ryan D, Obied HK, Arienzo M. Adsorption of phenols from olive oil waste waters on layered double hydroxide, hydroxyaluminium–iron-co-precipitate and hydroxyaluminium–iron–montmorillonite complex. Applied Clay Science. 2013;80-81:154–161. DOI:10.1016/j.clay.2013.01.014

[47] Tsipas SA. Effect of dopants on the phase stability of zirconia-based plasma sprayed thermal barrier coatings. Journal of the European Ceramic Society. 2010;30:61–72. DOI: 10.1016/j.jeurceramsoc.2009.08.008

[48] Wiwattanapongpan J, Mekasuwandumrong O, Chaisuk C, Praserthdam P. Effect of dopants on the properties of metal-doped zirconia prepared by the glycothermal method. Ceramics International. 2007;33:1469–1473. DOI: 10.1016/j.ceramint.2006.05.014

[49] Trunschke A, Hoang DL, Radnik J, Lieske H. Influence of lanthana on the nature of surface chromium species in la2o3-modified crox/zro$_2$ catalysts. Journal of Catalysis. 2000;191:456–466. DOI:10.1006/jcat.1999.2791

[50] Ramos MS, Santos MS, Gomes LP, Albornoz A, Rangel MC. The influence of dopants on the catalytic activity of hematite in the ethylbenzene dehydrogenation. Applied Catalysis A. 2008;341:12–17. DOI:10.1016/j.apcata.2007.12.035

Review of Recent Developments in CO$_2$ Capture Using Solid Materials: Metal Organic Frameworks (MOFs)

Mohanned Mohamedali, Devjyoti Nath, Hussameldin Ibrahim and Amr Henni

Additional information is available at the end of the chapter

Abstract

In this report, the adsorption of CO$_2$ on metal organic frameworks (MOFs) is comprehensively reviewed. In Section 1, the problems caused by greenhouse gas emissions are addressed, and different technologies used in CO$_2$ capture are briefly introduced. The aim of this chapter is to provide a comprehensive overview of CO$_2$ adsorption on solid materials with special focus on an emerging class of materials called metal organic frameworks owing to their unique characteristics comprising extraordinary surface areas, high porosity, and the readiness for systematic tailoring of their porous structure. Recent literature on CO$_2$ capture using MOFs is reviewed, and the assessment of CO$_2$ uptake, selectivity, and heat of adsorption of different MOFs is summarized, particularly the performance at low pressures which is relevant to post-combustion capture applications. Different strategies employed to improve the performance of MOFs are summarized along with major challenges facing the application of MOFs in CO$_2$ capture. The last part of this chapter is dedicated to current trends and issues, and new technologies needed to be addressed before MOFs can be used in commercial scales.

Keywords: CO$_2$ capture, solid sorbent, MOFs, ZIFs

1. Introduction

1.1. Environmental problem and CO$_2$ emissions

The increasing level of CO$_2$ emission is considered one of the major environmental challenges that our planet is facing today. The concentration of greenhouse gases in the atmosphere reached a new record in 2013, with CO$_2$ at 396 ppm which represents 142% of the concentration of the pre-industrial era [1]. Findings of a recent global atmosphere watch reported in a greenhouse gas bulletin [1] revealed that CO_2 concentration has increased between 2012 and 2013, more than any other year since 1984, which was attributed to the reduced uptake by the earth's biosphere. This alarming level of CO_2 shows the urgency for taking immediate actions to prevent serious repercussions of climate change. On December 2015, at the Paris Climate Conference (COP21), 195 countries adopted a historical and the first legally binding global climate agreement to keep the increase in global average temperature to well below 2°C above pre-industrial levels. The discovery of new fossil fuel reserves, combined with rising energy demand, led to an increase in the number and capacities of power plants worldwide. This situation is expected to extend into the future due to various factors such as industrial development and economic growth, especially in developing nations, which in turn is expected to further contribute to increasing levels of greenhouse gas emissions in years to come. According to a recent report by the Energy Information Administration, energy consumption is projected to rise by 56% between 2010 and 2040. Fossil fuels will continue to supply about 80% of the world energy through 2040. Industrial energy consumption represents the greatest share of emissions and is projected to consume more than 50% of the energy delivered in 2040. According to currently implemented regulations regarding fossil fuels, CO_2 emissions from power plants is projected to increase by 46% compared to emission level in 2010 [2].

Among several approaches that could be used to overcome the greenhouse gas effect is the utilization of clean energy alternatives which could be the ultimate solution to the climate change problem in terms of reducing CO_2 emissions. However, these green technologies still require significant modifications to the current energy framework. The great challenges facing these green technologies lie in the difficulty for implementation at industrial scale, which makes it economically infeasible when compared to fossil fuel-based power plants. This implies that unless green energy alternatives and energy infrastructure for the commercialization and the implementation of these new technologies are attained, the pursuit of new CO_2 emission reduction technologies will continue to be the most practical method to address greenhouse gas effects until the advancement in clean energy technologies reaches commercial stages.

There are three different strategies to reduce emissions of CO_2 from fossil fuel-based power plants. These include post-combustion capture in which CO_2 is separated from the combustion flue gas stream that is mainly composed of nitrogen and some other minor components such as water vapor and oxygen. The separation process in this scheme is a downstream unit which allows for an easy retrofit of a post-combustion capture unit to an existing power plant. However, the limitations of this technique include a low CO_2 partial pressure, relatively high flue gas temperature and large quantities of CO_2 in the flue gas stream [3, 4]. In the pre-combustion capture scenario, the fossil fuel is treated under certain temperature and pressure to gasify the fuel and produce hydrogen. This method offers streams with high CO_2 partial pressure and thus easy separation by utilizing variety of solvents; however, it requires significant modifications to the power generation plant. The last scenario is called the oxy-fuel capture in which the fuel is burned under a pure oxygen environment which requires the separation of oxygen/nitrogen from an air stream. The process produces pure CO_2 and water vapor which can be easily recovered through a simple condensation unit. Each separation scenario requires a different capture technology, and therefore the properties, characteristics, and operation of the separation process are also entirely different among the three strategies. The most advanced process for implementation in the field is post-combustion. We will therefore, in this chapter, focus on the post-combustion separation applications.

1.2. Existing technologies for CO_2 capture

In order to locate metal organic frameworks (MOFs) on the map of the technologies used for CO_2 capture applications, we will briefly describe the major technologies that have been employed and discuss their advantages and limitations. Figure 1 shows the different technologies used for CO_2 capture, whereas MOFs are used under the category of membranes and adsorbents.

The most widely investigated technology for CO_2 capture from flue gas is absorption using aqueous amine solutions such as monoethanolamine (MEA), diethanolamine (DEA), and methyldiethanolamine (MDEA), as well as blends of different amines [5–7]. Amine scrubbers are considered a well-developed technology and is available in commercial scale for post-combustion capture applications [8]. The major limitations of this technology include the high energy required for solvent regeneration, stability of the amine system at the regeneration conditions, and the negative influence of impurities present in the flue gas that might significantly affect the stability and performance of the solvent [9, 10].

Under the category of absorption technology and in order to overcome the limitations of amine-based technologies, aqueous ammonia as a solvent for CO_2 separation has also been widely used to benefit from the low heat of absorption of ammonia-based solvents as compared to amine systems. Besides, the ammonia can also absorb other impurities existing in the gas stream such as NO and SO_x. The major drawback of ammonia-based solvents lies in the need to cool the flue gas prior to introducing it to the absorption column to prevent ammonia losses to the gas stream. This adds a huge energy requirement considering the large volume of flue gas stream that typically needs to be treated [11]. The chilled ammonia process faces similar issues in addition to fouling of the heat exchanger by ammonium bicarbonate deposition from saturated solutions [4].

Great efforts have been made to find new and efficient materials for absorptive CO_2 separation. Ionic liquids (ILs) are liquid salts composed of cations and anions, have been proposed as promising solvents to replace the existing amine-based solvents. ILs possess several remarkable properties that make their application in CO_2 separation one of the hottest research topics in the last few years [12–14]. These properties include low volatility, high CO_2 solubility, good thermal stability, and the possibility of systematically tuning the structure toward certain properties [15–17]. Several review papers reporting experimental data related to CO_2 solubility, selectivity, effect of ILs structure on performance, and the stability of ILs are available [12, 18]. Recent developments on the application of the amine-modified ILs, known as task-specific ILs (TSILs), are also widely discussed in the literature [19, 20], including both physical and chemical interactions with CO_2. Unfortunately, many ILs and TSILs suffer from a common problem of high viscosity after CO_2 absorption. Even though some recent reports mentioned the availability of ILs with low viscosities, it is still evident that much work has to be done to overcome this limitation. Finding cheap routes for the synthesis of these materials is one of the greatest challenges facing researchers working in this area [21]. In this chapter, a great portion will be dedicated to the incorporation of ILs into the pores of MOFs to improve their CO_2 capture capabilities.

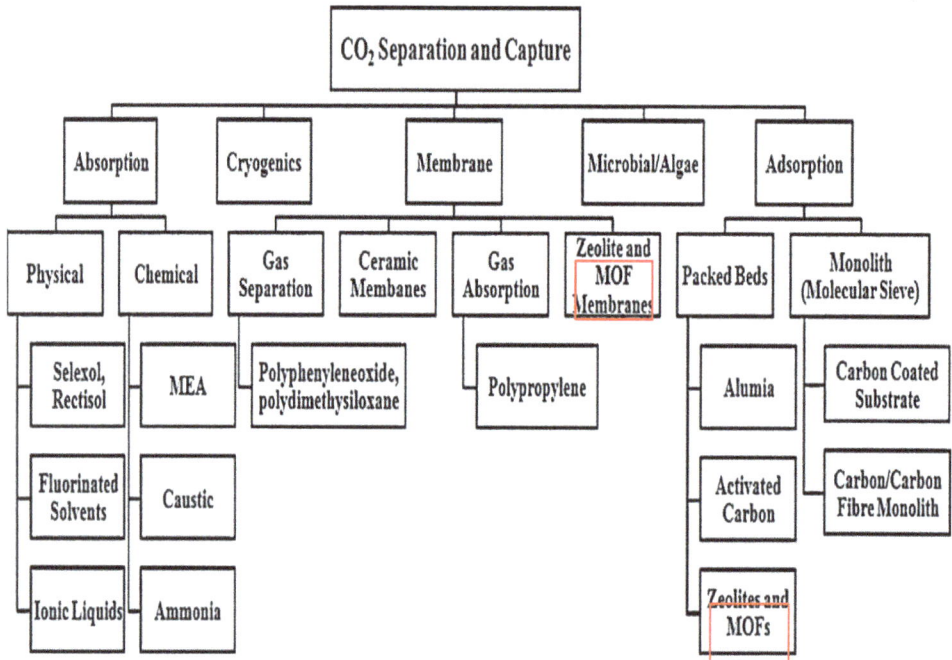

Figure 1. Different technologies used for CO_2 capture [22].

2. CO_2 capture using solid sorbents

2.1. Criteria for the evaluation of solid sorbents

In order to evaluate solid materials for their performance in CO_2 separation from flue gases, some important performance criteria must be met. These include:

- **Adsorption capacity**: it is a key criteria in evaluating solid sorbent performance. It provides information on the amount of CO_2 that could be adsorbed by a given solid material. It can be represented in terms of gravimetric uptake which is the amount of CO_2 adsorbed per unit mass sorbent (gram CO_2/gram sorbent, or cm^3 CO_2/gram sorbent). The volumetric uptake is another measure for capacity, and it reports the CO_2 uptake per volume of sorbent material (gram CO_2/cm^3 sorbent, or cm^3 CO_2/cm^3 sorbent). This criterion is of great importance because it represents the amount of sorbent needed for a particular duty and therefore the size of the adsorption bed. It is also considered a crucial factor in determining the energy requirement during the regeneration step.

- **Selectivity for CO_2:** it represents the CO_2 uptake ratio to the adsorption of any other gas (typically nitrogen for post-combustion capture, and methane for natural gas). It is an essential evaluation criterion, and affects the purity of the adsorbed gas, which will significantly influence the sequestration of CO_2. The simplest method to estimate the selectivity factor is to use single-component adsorption isotherms of CO_2 and nitrogen.

- **Enthalpy of adsorption:** it represents another critical parameter in the evaluation of the performance of solid sorbents. It is a measure of the energy required to regenerate the solid sorbent, and it therefore significantly influences the cost of the regeneration process. It represents the affinity of the material toward CO_2 and measures the strength of the adsorbate–adsorbent interaction.

- **Physical, thermal, and chemical stability:** in order to reduce operating costs, solid sorbents must demonstrate stability under flue gas conditions, adsorption operation conditions, and during the multi-cycle adsorption–regeneration process. In particular, stability in the presence of water vapor is essential for the sustainable performance of the solid sorbent. In addition to thermal properties of the solid sorbent, heat capacity and thermal conductivity are also important in heat transfer operations.

- **Adsorption/desorption kinetics:** the time of the adsorption–regeneration cycle greatly depends on the kinetics of the CO_2 adsorption–desorption profile, which is measured in breakthrough experiments. Sorbents that adsorb and desorb CO_2 in a shorter time are preferred as these reduce the cycle time as well as the amount of sorbent required, and ultimately the cost of CO_2 separation.

- **Cost of the sorbent material:** it is an important factor in the selection of the sorbent material. Materials that exhibit excellent adsorption attributes, and are readily available at low cost, are considered the main targets for researchers working in the field of CO_2 capture. Besides, the environmental impact of synthesizing these materials is considered one of the greatest challenges to overcome.

In the following sections, we describe the main solid sorbents used for CO_2 capture, their applications, major attributes, and limitations.

2.2. Zeolites

Zeolites are porous crystalline aluminosilicate materials available naturally, but can also be prepared synthetically. The zeolite framework is composed of tetrahedral T atoms where T could be Si or Al, connected by oxygen atoms to form rings of different pore structures and sizes. The pore size of the zeolite framework varies between 5 and 12 Å [23]. They are widely used as catalysts in the refining industry [24, 25], fine chemicals synthesis [26, 27], and in gas separation applications [28, 29]. Zeolites are considered promising candidates in CO_2 capture application as has been widely reported in the literature [30–32]. CO_2 can be adsorbed on zeolites through different mechanisms, such as molecular sieving effect based on the difference in size [33, 34]. Separation can also take place based on polarization interactions between the gas molecule and the electric field on the charged cations in the zeolite framework [33]. Accordingly, CO_2 removal with zeolites can be controlled by changing the pore size, polarity, and the nature of the extra framework cation. Among the different zeolites investigated for CO_2 capture applications, zeolite 13X is the most widely studied sorbent, and is considered the benchmark technology for solid sorbents [35, 36]. Research on the use of zeolites as sorbents for CO_2 capture can be categorized into different areas depending on the approach and the techniques adopted to address the improvement in capture performance. These categories comprise tuning the pore size, designing zeolites with controlled polarities, investigating novel zeolites, optimizing the cation exchange, and most recently incorporating amine moieties and other chemical functions into the zeolite frameworks. Ocean et al. [37] have studied the selectivity to adsorb CO_2 by controlling the pore size of an NaKA zeolite through the synthesis of nanosized NaKA zeolites. Overall, the adsorption kinetics on the nanosized crystals was fast enough for CO_2 capture applications; however, the formation of a thin layer on the nanosized NaKA zeolite, due to intergrowth on the surface, did not considerably improve the adsorption kinetics. In contrast, Goj et al. [38] performed atomistic simulations for silicalite, ITQ-3, and ITQ-7, and reported a positive effect on CO_2 uptake and selectivity by tuning the pore apertures. Sravanthi et al. [39] provided a novel approach to control the pore size and volume by utilizing pore expansion agents and obtained average pore size around 30 nm. The application of the pore-expanded MCM-41 in CO_2 separation resulted in the uptake of about 1.2 mmol/g.

Several studies have been conducted to control zeolite affinity toward CO_2, which can be realized by tuning the polarity of the zeolite through alteration of the Si/Al ratio and the nature of the cation. Remy et al. [33] studied the selective separation of CO_2 on low-silica KFI zeolite (Si/Al = 1.67) by employing ion exchange with Na, Li, and K. Li-exchanged KFI has shown the highest CO_2 uptake which was attributed to the large pore volume as compared to Na and K cations. In comparison with high-silica KFI sample (Si/Al = 3.57–3.67), Li–KFI had the highest capacity at low pressure due to the strong electrostatic field. The overlap between pore size and polarity effects is also strongly observed for amine-supported zeolites, which have gained considerable attention in the last few years [40–45]. For instance, Ahmad et al. [46] have impregnated melamine into β-zeolite and obtained dynamic CO_2 uptake of 3.7 mmol/g at atmospheric pressure and temperature of 25 °C. The major challenge facing amine-modified zeolites is the tradeoff between the increased affinity toward CO_2 (strong interaction with the sorbent) and the reduction in pore volume, and consequently the uptake, especially, at low pressures. Factors such as amines loading, distribution, and the nature of the cation can play a vital role to avoid the blockage of the porous structures with the bulky amine moieties [42, 47]. Kim et al. [48] have performed a rigorous investigation through the simulation of thousands of zeolites to evaluate the adsorption properties of these materials and identify the optimum structures for improved CO_2 separation attributes. This study provides a systematic approach to rank and select appropriate zeolites for the required capture objectives. However, important factors such as stability under humid environment, adsorbent and process cost, and the availability of zeolite structures were not taken into consideration.

The hydrophilic nature of most zeolite structures is considered a major drawback of zeolites especially for post-combustion CO_2 applications [49, 50]. Water competes with CO_2 on the available sorption sites and might influence the zeolite structure and framework [51]. As explained earlier, the presence of the exposed cation sites increases CO_2 uptake. In a recent study by Serena et al. [52], the relationship between the water content of the zeolite and the density of the cations was investigated, and a linear relationship was found to describe the decrease of the cation population with increasing water content. This observation highlights the detrimental effect of the presence of water vapor on the adsorption of CO_2 on zeolites.

2.3. Carbon-based CO_2 capture

Carbon-based adsorbents have been used for CO_2 separation in different forms including activated carbons (ACs), carbon nanotubes (CNTs), and graphenes. Activated carbons have an amorphous porous structure with high surface areas that are readily available for CO_2 uptake. They have been widely investigated as sorbents for CO_2 removal due to their low cost and the availability of raw materials [53–55]. However, there are no active sites to bond with the adsorbed CO_2 as cations in zeolite sorbents. This weak interaction results in lower enthalpy and therefore lower energy requirement for regeneration. On the contrary, ACs have very low CO_2 uptake at low pressures due to the absence of the electric field on the surface. Kacem et al. [56] performed a comparison between the performance of ACs and zeolite for CO_2/N_2 and CO_2/CH_4 separation based on their capacity, regeneration capacity, and reusability. It was concluded that at higher pressures (above 4 bars), the CO_2 uptake for ACs was much higher than zeolites. Also, the recovered CO_2 after the regeneration of ACs had higher purity than in the case of zeolites. When compared to zeolites, ACs maintain their adsorption stability even in the presence of water vapor which does not cause any framework failure [57].

In order to enhance the adsorption capacity on ACs, several studies have been conducted in order to improve the affinity toward CO_2 by introducing amine-based functional groups [58–61]. In a recent study, Maria et al. [62] described a systematic surface modification of micro-porous ACs through a stepwise chemical treatment. They were successful in grafting amine and amide functional groups on the surface of ACs with only 20% loss of surface area. Gibson et al. [63] studied the polyamine-impregnated porous carbons and achieved 12 times higher CO_2 capacity than bare porous carbon. Chitosan and triethylenetetramine have been success-fully impregnated onto the surface of ACs and have shown 60 and 90% increased CO_2 uptake at 298 K and 40 bars. In addition to amine functional groups, ammonia-modified ACs, at atmospheric pressure and a temperature range from (303 to 333) K, have been studied [64]. Authors report that an enthalpy of 70.5 kJ/mol was obtained compared to 25.5 kJ/mol for the pristine ACs, suggesting the possibility of chemisorption. Another report has also supported the improved adsorption capacity and selectivity by employing NH_3 at high temperature and has considerably improved CO_2 uptake from 2.9 mmol/g for the bare AC to 3.22 mmol/g for the modified one at 303 K and 1 bar.

Several studies have been dedicated to the application of amine-modified carbon nano tubes (CNTs) as solid sorbents for CO_2 separation [65–69]. Industrial grade CNTs have been functionalized with tetraethylenepentamine (TEPA) by Liu et al. [65], and the effects of amine loadings on the CO_2 uptake, heat of adsorption, and adsorbent regenerability were investigated. TEPA-impregnated CNTs have shown an enhanced capacity of 3.09 mmol/g at 343 K. Similar studies were also reported using different amines such as (3-aminopropyl)triethoxysilane (APTES) [70], polyethyleneimine (PEI) [67], and other amines (primary, secondary, tertiary, diamines, and tri-amines) [71].

Graphene is a planar sheet of carbon atoms extended in two dimensions, and was discovered in 2004 [72]. Graphite-based capture was recently introduced (after 2011) as a promising candidate for CO_2 capture applications, and research is growing rapidly in this area [73–77]. A recent review by Najafabadi is available on the current status and research trends of using graphene and its derivatives as solid sorbents for CO_2 capture [78]. Research in this area involves grafting various functional groups on graphene such as N-doped graphene composites (surface area = 1336 m^2/g), as reported by Kemp et al. [79], which showed a reversible CO_2 capacity of 2.7 mmol/g at 298 K and 1 atm as well as enhanced stability for repeated adsorption cycles. Borane-modified graphene was also reported by Oh et al. [80], obtaining a CO_2 uptake of 1.82 mmol/g at 298 K and 1 atm. Some novel hybrid materials have also been introduced to obtain better improvements in the adsorption properties, including mesoporous graphene oxide (GO)-ZnO nanocomposite [81], mesoporous TiO_2/graphene oxide nanocomposites [82], Mg–Al layered double hydroxide (LDH), graphene oxide [83], MOF-5 and aminated graphite oxide (AGO) [84], UiO-66/graphene oxide composites [85], and MIL-53(Al) and its hybrid composite with graphene nanoplates (GNP) [86].

2.4. Metal organic frameworks

A more recent class of porous materials was manufactured and named metal organic frameworks. They represent one of the promising adsorbents and have gained significant attention during recent years for gas separation applications [87, 88]. MOFs are composed of metal ions or clusters (nodes) bridged by organic ligands (connecters) to form various structures and networks. MOFs are well recognized for their extraordinary surface areas, ultrahigh porosity, and most importantly the flexibility to tune the porous structure as well as the surface functionality due to the presence of organic ligands that can easily be chemically modified [89, 90]. One main advantage of MOFs over other solid materials is the possibility to tailor the pore size and functionality by rational selection of the organic ligand, functional group, metal ion, and activation method.

Several review papers are available in the literature for gas separation using MOFs [91–96]; however, great progress has been achieved during the past four years (2012 onward). In order to address the limitations of MOFs and investigate new structures, novel functional groups, in addition to hybrid systems and technologies, more studies are needed to explore the mechanisms involved and to improve the uptake capacity in a humid environment. For these reasons, considerable effort has been observed during the past decade to address gas separation and adsorption using MOFs. Figure 2 shows the number of publications on CO_2 capture and separation using MOFs during the past 15 years, which reflects the growing interest of MOFs as efficient solid sorbents.

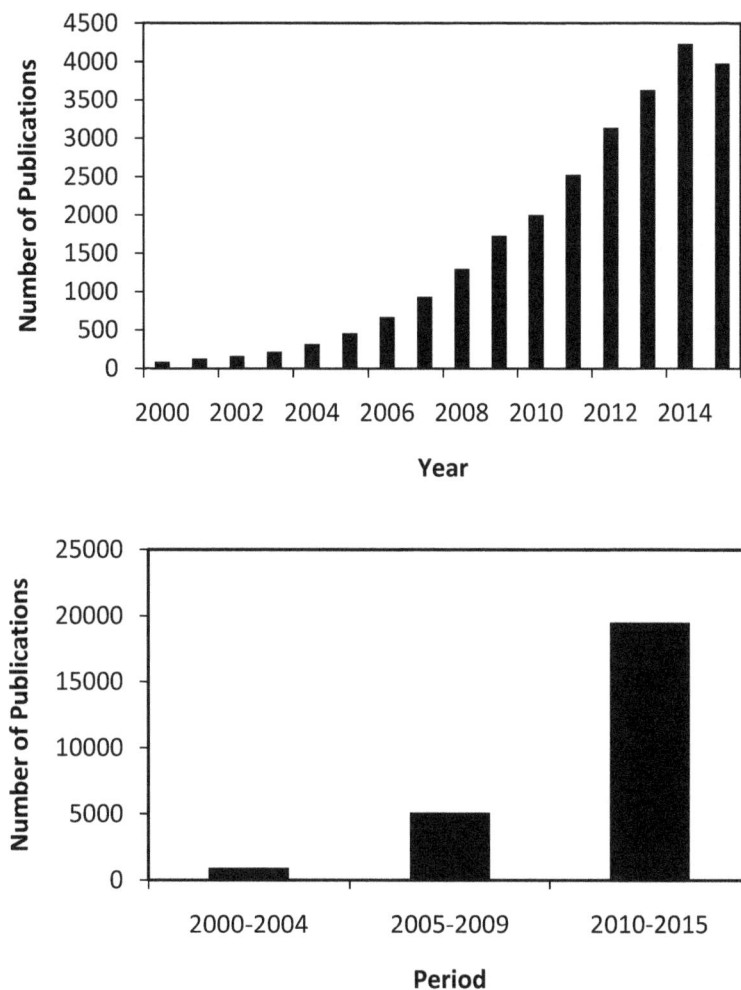

Figure 2. Number of publications on CO_2 capture using MOFs (based on Web of Science database)

3. Adsorption of CO_2 on metal organic frameworks

CO_2 capture performance of different MOFs will be comprehensively reviewed in terms of their capacity, selectivity, heat of reaction, and major challenges facing researchers, and some ideas to approach these challenges will also be provided. The next section is dedicated to review the most recent studies of CO_2 capture and separation on MOFs, and we will mainly target the works published in the last four years.

3.1. Evaluation of MOFs in CO_2 Capture

As introduced earlier, capacity, selectivity, and heat of adsorption are considered the main criteria for the evaluation of MOFs for CO_2 separation. CO_2 uptake is a proportional function of pressure in the gas phase, where the low pressure corresponds to post-combustion applications. The gravimetric uptake of CO_2 is indicative of the ability of MOFs to adsorb CO_2 and, therefore, we have reported CO_2 uptake along with MOF surface area, and other properties for MOFs published after 2012 which could be added to the published reviews that have listed these data in a table format. Table 1 represents the properties of MOFs at high-pressure applications, while Table 2 presents the low-pressure data.

	Surface Area (m²/g)							
Common name	BET	Langmuir	Capacity (wt%)	Pressure (bar)	Temp. (K)	Selectivity	Qst (kJ/mol)	Ref.
UiO(bpdc)	2646	2965	72.5	20	303			[97]
ZJU-32	3831		49	40	300			[98]
UPG-1	410	514	11.9	9.8	298	24	24	[99]
Cu₃(H₂L²)(bipy)₂.11H₂O			6.4	8.5	298			[100]
Cu₃(H₂L²)(etbipy)₂.24H₂O			4.7	9.6	298			[100]
NU-111	4932		61.8	30	298		**23**	[101]
HTS-MIL-101	3482		52.8	40	298			[102]
DGC-MIL-101	4198		59.8	40	298			[102]
UTSA-62a	2190		43.7	55	298		16	[103]
ZIF-7	312	355	20.9	10	298		33	[104]
{Ag3[Ag5(l3-3,5-Ph2tz)6](NO3)2}n			12.3	10	298	10.5	19.1	[105]
{Ag3[Ag5(l3-3,5-tBu2tz)6](BF4)2}n			5.4	10	298	14	15	[105]
Basolite® C 300	1706.42		41.9	224.99	318		18	[106]
Basolite® F300	1716.46		24.1	224.99	318		19	[106]
Basolite® A100	1524.8		26.9	224.99	318		9	[106]
IRMOF-8	1599	1801	7.8	1	298		21.1	[107]
IRMOF-8-NO2	832	926	3.8	1	298		35.4	[107]
MIL-101(Cr)	2549		24.2	30	303			[108]
HKUST-1	1326		26.3	30	303			[108]
DMOF	1980		38.1	20	298	12[a]	20	[109]
DMOF-DM1/2	1500		27.5	20	298			[109]
DMOF-Br	1320		24.3	20	298			[109]
DMOF-NO2	1310		32	20	298			[109]

Common name	Surface Area (m²/g)		Capacity (wt%)	Pressure (bar)	Temp. (K)	Selectivity	Qst (kJ/mol)	Ref.
	BET	Langmuir						
DMOF-TM1/2	1210		23.9	20	298			[109]
DMOF-TF	1210		16.2	20	298	9[a]	18	[109]
DMOF-Cl2	1180		26.4	20	298	17	21	[109]
DMOF-OH	1130		24.8	20	298			[109]
DMOF-DM	1120		25.4	20	298	23[a]	23	[109]
DMOF-TM	1050		23.6	20	298	28[a]	29	[109]
DMOF-A	760		17.1	20	298			[109]

[a] IAST selectivity

Table 1. Adsorption capacities at high pressure

Common name	Surface Area (m²/g)		Capacity (wt%)	Pressure (bar)	Temp. (K)	Selectivity	Qst (kJ/mol)	Ref.
	BET	Langmuir						
rht-MOF-pyr	2100		12.7	1	298		28	[110]
rht-MOF-1	2100		11	1	298		29	[110]
JLU-Liu22	1487		15.6	1	298		30	[111]
SIFSIX-3-Zn			8.9	1	298			[112]
SIFSIX-3-Cu			9.6	1	298			[112]
SIFSIX-3-Co	223		10	1	298		47	[112]
SIFSIX-3-Ni	368		10.3	1	298		51	[112]
{[H$_2$N(CH$_3$)$_2$]$_4$[Zn$_9$O$_2$(BTC)$_6$(H$_2$O)$_3$].3DMA}c$_n$	844	1132	10.9	0.91	298		29	[113]
{[NH$_2$(CH$_3$)$_2$, Cd(BTC)].DMA}$_n$	406	539	6.4	0.91	298	30	34.7	[113]
Ni-DOBDC	798		18.2	1	298			[114]
Py-Ni-DOBDC	409		12	1	298			[114]
UiO(bpdc)	2646	2965	8	1	303			[97]
ZJU-32	3831		4.8	1	300			[98]
Cu-TDPAH	1762		18.4	1	298	200[a]	33.8	[115]
Zn/Ni-ZIF-8-1000		750	9.9	1	298	30[a]	61.2	[116]
ZIF-8-1000			9.6	1	298	23.5[a]	49.7	[116]

Common name	Surface Area (m²/g)		Capacity (wt%)	Pressure (bar)	Temp. (K)	Selectivity	Qst (kJ/mol)	Ref.
	BET	Langmuir						
Zn(5-mtz)(2-eim).(guest) [ZTIF-1]	1430	1981	8.2	1	295	81	22.5	[117]
Zn(5-mtz)(2-pim).(guest) [ZTIF-2]	1287	1461	3.8	1	295		20	[117]
UTSA-49	710.5	1046.6	13.6	1	298	95.8		[118]
ZJNU-40	2209		16.4	1.01	296		18.4	[119]
UPG-1	410	514	2.1	1	298	24	24	[99]
InOF-8			6.9	1	295	45.2		[120]
$Cu_3(H_2L^1)(bipy)_2.9H_2O$			2.5	1	195			[100]
$Cu_3(H_2L^2)(bipy)_2.11H_2O$			2.3	1	298			[100]
$Cu_3(H_2L^2)(etbipy)_2.24H_2O$			0.5	1	298			[100]
UiO-66(Zr100)	1390	1644	6.2	1	298		26	[121]
UiO-66(Ti32)	1418	1703	6.4	1	298		28	[121]
UiO-66(Ti44)	1749	2088	7.2	1	298		34	[121]
UiO-66(Ti56)	1844	2200	8.8	1	298		37	[121]
NU-111	4932		4.8	1	298		23	[101]
JLU-Liu1	145	221	5.9	1	298		47.7	[122]
HTS-MIL-101	3482		12.3	1	298			[102]
DGC-MIL-101	4164		14.5	1	298			[102]
UNLPF-1			13.9	1	273			[123]
UTSA-62a	2190		8.1	1	298		16	[103]
[Zn2(BME-bdc)x(DB-bdc)2_xdabco]n			21.7	0.91	195			[124]
Zn-DABCO	1870	1902	7.2	1	298		22.4	[125]
Ni-DABCO	2120	2219	8.1	1	298		25.8	[125]
Cu-DABCO	1616	1678	6.2	1	298		22.4	[125]
Co-DABCO	2022	2095	4.1	1	298		29.8	[125]
ZnAcBPDC	920		11.7	0.9	293			[126]
ZnBuBPDC	850		7.6	0.89	293			[126]
Mg/DOBDC	1415.1		25	1	298		47	[127]
Co/DOBDC	1089.3		21.6	1	298		37	[127]

	Surface Area (m^2/g)		Capacity (wt%)	Pressure (bar)	Temp. (K)	Selectivity	Qst (kJ/mol)	Ref.
Common name	BET	Langmuir						
Ni/DOBDC	1017.5		20.5	1	298		42	[127]
MIL-100(Cr)	1528.7		9.5	1	298			[127]
ZIF-7	312	355	9.1	1	298			[104]
{Ag3[Ag5(l3-3,5-Ph2tz)6](NO3)2}n			1.6	1	298	10.5	19.1	[105]
{Ag3[Ag5(l3-3,5-tBu2tz)6](BF4)2}n			1.6	1	298	14	15	[105]
CuBTTri	1700		10.8	1	293			[128]
pip-CuBTTri	380		7.1	1	293	130[a]	96.5	[128]
Basolite® C 300	1706.42		9.4	0.95	318		18	[106]
Basolite® F300	1716.46		2.4	0.95	318		19	[106]
Basolite® A100	1524.8		4.4	0.95	318		9	[106]
IRMOF-8	1599	1801	51.2	30	298		21.1	[107]
IRMOF-8-NO2	832	926	31.3	30	298		35.4	[107]
CPM-5	2187		8.8	1	298	16.1	36.1	[129]
Ni-MOF-74	1252	1841	19.4	1	298			[130]
Mg-MOF-74	1416	2085	30.1	1	298			[130]
MIL-101(Cr)	2549		6.8	1	303			[108]
HKUST-1	1326		13.2	1	303			[108]
[Cu(tba)2]n			7.3	1	293	25[a]	36.0	[131]
IRMOF-74-III-CH3	2640		10	1	298			[132]
IRMOF-74-III -NH2	2720		10.4	1	298			[132]
IRMOF-74-III-CH2NHBoc	2170		7	1	298			[132]
IRMOF-74-III-CH2NMeBoc	2220		6.6	1	298			[132]
IRMOF-74-III-CH2NH2	2310		10.8	1	298			[132]
IRMOF-74-III-CH2NHMe	2250		9.6	1	298			[132]
DMOF	1980			1	298	12[a]	20	[109]
DMOF-DM1/2	1500		8.1	1	298			[109]
DMOF-Br	1320		6.4	1	298			[109]

Common name	Surface Area (m²/g)		Capacity (wt%)	Pressure (bar)	Temp. (K)	Selectivity	Qst (kJ/mol)	Ref.
	BET	Langmuir						
DMOF-NO2	1310		9.9	1	298			[109]
DMOF-TM1/2	1210		8.1	1	298			[109]
DMOF-TF	1210		3.3	1	298	9[a]	18	[109]
DMOF-Cl2	1180		8.8	1	298	17[a]	21	[109]
DMOF-OH	1130		9.6	1	298			[109]
DMOF-DM	1120			1	298	23[a]	23	[109]
DMOF-TM	1050		13.3	1	298	28[a]	29	[109]
DMOF-A	760		10.6	1	298			[109]
CPM-33a	966	1257	12.6	1	298		22.5	[133]
CPM-33b	808	1119	19.9	1	298		25	[133]
Ni3OH(NH2bdc)3tpt	805	1115	14.8	1	298		21.5	[133]
Ni3OH(1,4-ndc)3tpt	222	310	4.6	1	298		25.3	[133]
Ni3OH(2,6-ndc)3tpt	1002	1392	7.9	1	298		24.7	[133]
Ni3OH(bpdc)3tpt	724	1009	5.5	1	298		18.7	[133]
ZIF-7-S	150		3.7	1	303			[134]
ZIF-7-D	25		9	1	303			[134]
ZIF-7-R	5		8.7	1	303		34	[134]
HKUST-1		2203	12.8	1	313			[135]
Fe-MIL-100		2990	6.6	1	313			[135]
Zn(pyrz)2(SiF6)			10.8	1	313			[135]
Mg2(dobpdc)		1940	23.8	1	313			[135]
Ni2(dobpdc)		1593	21.2	1	313			[135]
mmen-Mg2(dobpdc)			15.8	1	313			[135]
mmen-Ni2(dobpdc)			7.3	1	313			[135]
mmen-CuBTTri			11.3	1	313			[135]

[a] IAST selectivity

Table 2. Adsorption capacities at low pressure

3.2. Strategies to Improve the CO_2 Capture Performance on MOFs

Several strategies have been adopted to improve the performance of MOFs in CO_2 capture applications. The ability to precisely tune the MOF structures has led to versatile approaches that can be utilized to enhance CO_2 uptake, selectivity, and the affinity toward CO_2. These methods could be classified into effects of open metal sites, pre-synthetic modifications of the organic ligand, and post-synthetic functionalization schemes.

3.2.1. Open Metal Sites

Open metal sites in MOFs are formed by the removal of a solvent molecule coordinated to the metal nodes by applying vacuum and/or heat after the synthesis of framework in a process called "activation." The presence of open metal sites on the MOF framework has a great impact on the selectivity toward CO_2 as well as on the binding energy between the adsorbed CO_2 molecules and the surface of MOF sorbents. These coordinately open metal centers act as binding sites where CO_2 molecules can attach and bind to the pore surface by the induction of dipole–quadrupole interactions. Allison et al. [136] have developed a systematic procedure to precisely understand the interactions between the CO_2 molecule and the force field generated by the open metal sites in MOF-74. The developed method allows for accurate estimation of adsorption isotherms using computational approach which enables the evaluation of different hypothetical open metal sites. These observations confirm previous findings of Kong et al. on understanding CO_2 dynamics in MOFs with open metal centers [137]. Among the MOF family, HKUST-1, M-MIL-100, M-MIL-101, and M-MOF-74 are the most widely studied frameworks with open metal sites (M represents the metal site). However, to precisely investigate the influence of the open metal sites, we need to isolate the effects of the nature of organic ligands, the synthesis route, and functional groups present in the framework. It was observed that utilizing light metal sites provides higher surface areas, and therefore improve CO_2 uptake at low pressures for MOF-74 [138]. Several studies have reported the effects of metal centers using computational approach as reported for M-MOF-74 [138–140] where noble metals such as Rh, Pd, Os, Ir, and Pt are considered promising candidates for CO_2 capture (see Figure 3).

Casey et al. [141] studied the isostructural series of HKUST-1 for various metal centers (Mo, Ni, Zn, Fe, Cu, and Cr) to get insights into the adsorption mechanism and the force field created by different metal types. It was found that the presence of divalent metals such as Mg^{2+} significantly increased CO_2 binding strength and resulted in higher selectivity toward CO_2. In addition to the nature of the metal nodes, it was found that the activation method plays a vital role in determining CO_2 uptake and affinity toward CO_2 which was in agreement with Llewellyn et al. [142] for MIL-100 and MIL-101, where various activation methods resulted in different CO_2 loadings and heat of adsorption.

In a recent study, Cabelo and coworkers [143] investigated the interaction between CO_2 and the unsaturated Cr(III), V(III), and Sc(III) metal sites in MIL-100 framework using variable temperature infrared spectroscopy. The enthalpy of adsorption for Cr(III), V(III), and Sc(III) were amounted to be (−63, −54, and −48) kJ/mol, respectively, which are considered among the highest values for CO_2 adsorption on MOFs with open metal centers to date. The synthesis and characterization of an M-DABCO series (M = Ni, Co, Cu, Zn) were described by Sumboon et al. [125] to systematically evaluate the effect of the metal identity on surface area, pore volume, and CO_2 uptake. It was concluded that Ni-DABCO has shown the highest pore volume and specific surface area due to the high charge density present at the metal center. Comparison

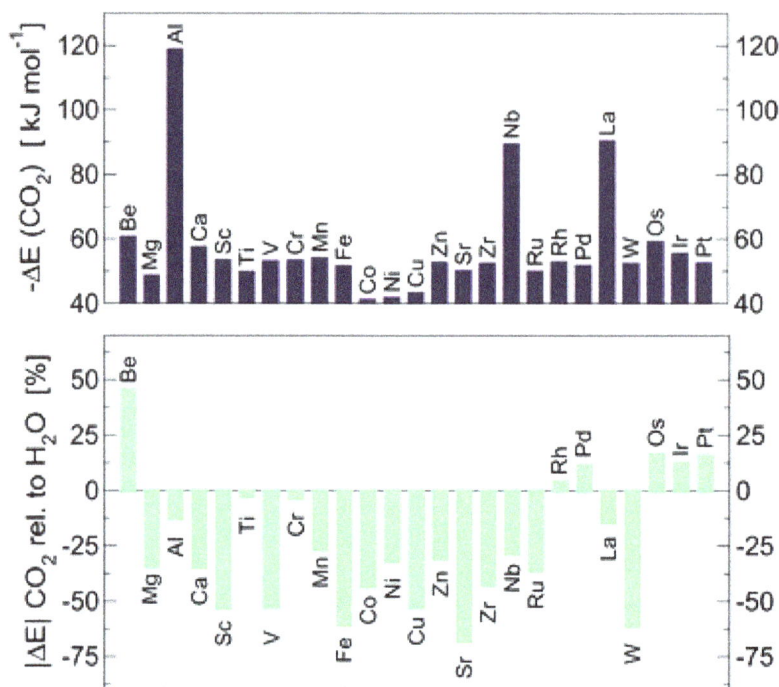

Figure 3. Top: ΔE for CO_2 adsorption (in kJ/mol) in M-MOF-74. Bottom: Magnitude of the adsorption energy of CO_2 relative to H_2O. A positive value in this plot means that CO_2 binds more strongly than H_2O (Adapted from [139]).

of the M-DABCO with activated carbons and MIL-100(Cr) revealed that the unsaturated cations possess exceptional CO_2 uptake of 180 cm³/g at 1 bar and 298 K [as compared to 30 cm³/g for ACs and 60 cm³/g for MIL-100(Cr),144].

3.2.2. Pre-synthetic Modifications of MOFs

Organic ligands are the linkers that connect the metal nodes together and therefore determine the framework structure, pore volume, pore window, and surface area which are very crucial characteristics in CO_2 separation process. Ligand functionalization is considered to be a powerful tool to improve the adsorption of CO_2 on MOFs due to the wide range of functional groups and the ease of modifying the organic ligand through strong covalent interactions. In a recent computational work by Torrissi et al., the impacts of various functional groups attached to the ligand part were investigated by density functional theory (DFT) [145]. The incorporation of amine functional moieties to the organic ligands has witnessed much attention in recent years, due to the proven positive effect of the presence of open nitrogen sites on the MOF frameworks [146]. Keceli et al. [147] studied four biphenyl ligands modified with amide groups of different chain lengths. Varying the length of the alkylamide group has shown a great impact on the porosity, surface area, and CO_2 capacity. It was also evident that the activation procedure has great influence on the surface area of the resulting material which is attributed to the different mechanisms of solvent removal from the MOF framework. Three amino-functionalized MOFs have been prepared from 2-aminoterephthalate (ABDC) and three different metals (Mg, Co, and Sr). Despite a low surface area (63, 71, and 2.5 for Mg, Co, and Sr, respectively) and a relatively low CO_2 uptake (1.4 mmol/g at 1 bar and 298 K), the prepared MOFs had exceptional selectivity toward CO_2 (396 was recorded for Mg-ABDC) and exhibited high heat of adsorption [148]. Shimizu and coworkers [149] used 3-amino-1,2,4-triazole ligands to design a 3D structure MOF with 782 m^2/g surface area and 0.19 cm^3/g pore volume that is capable to achieve CO_2 uptake of 4.35 mmol/g at 1.2 bar and 273 K. Moreover, the as-synthesized MOF has shown enthalpy of adsorption of 40.8 kJ/mol at zero coverage which was comparable to the commercial zeolite NaX (48.2 kJ/mol). In a similar study, Xiong et al. [118] used triazole ligands to prepare a new framework called UTSA-49 by incorporating nitrogen atoms and methyl functional groups on 5-methyl-1H-tetrazole ligands which recorded 13.6 wt% CO_2 uptake at 1 bar and 298 K and 27 kJ/mol enthalpy (Figure 4). These observations were in agreement with work reported by Gao et al. for the influence of triazolate linkers [150]. It is essential to understand the synergistic effect between the multiple functional groups on the pore surface and their size exclusion effects which are considered potential approaches to optimize the performance of functionalized MOFs. Table 3 summarizes CO_2 capture properties of MOFs modified with different amino functional groups.

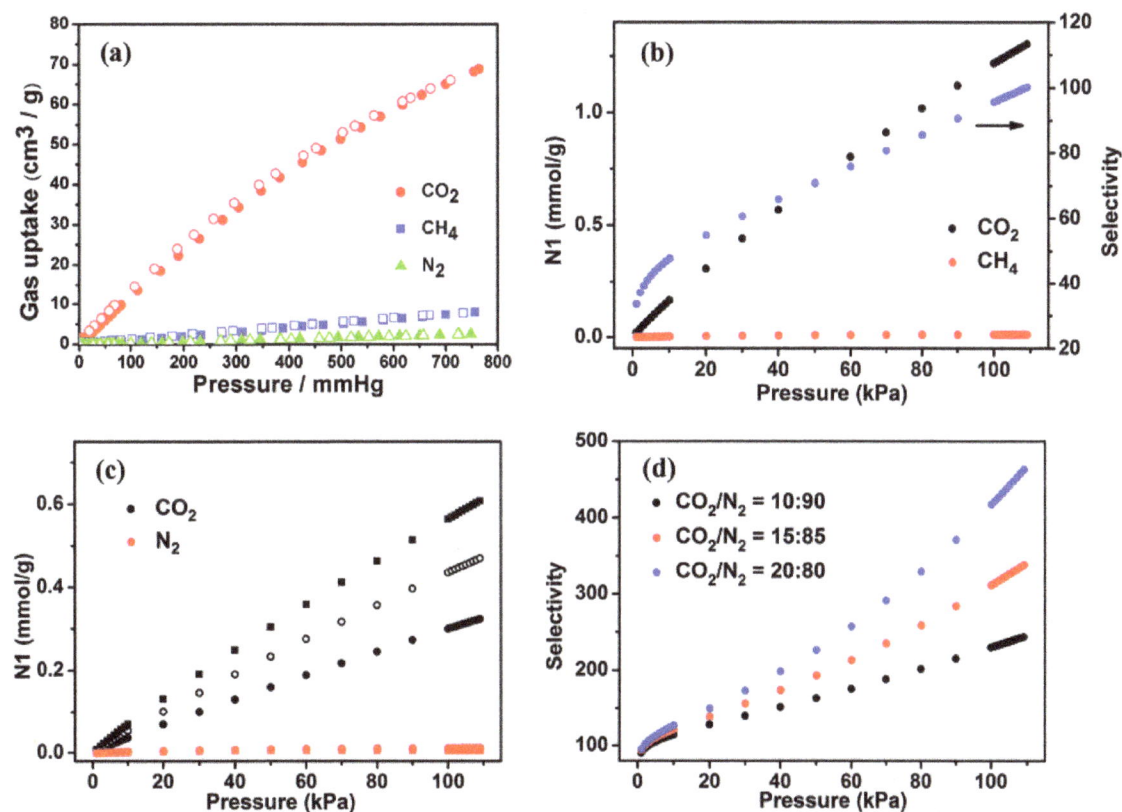

Figure 4. (a) Adsorption (solid) and desorption (open) isotherms of carbon dioxide (red circles), methane (blue squares), and nitrogen (green triangles) on UTSA-49a at 298 K. (b) Mixture adsorption isotherms and adsorption selectivity predicted by IAST of UTSA-49a for CO_2 (50%) and CH_4 (50%) at 298 K. (c) Mixture adsorption isotherms predicted by IAST of UTSA-49a for CO_2 and N_2 (10:90, 15:85, and 20:80) at 298 K. (d) Mixture selectivity predicted by IAST of UTSA-49a for CO_2 and N_2 (10:90, 15:85, and 20:80) at 298 K. Adapted from [118].

MOF name	Type of functional group	CO_2 uptake (wt. %)	Enthalpy of adsorption (kJ/mol)	Pressure	Temperature	Surface area (m^2/g)	Ref.
ZIF-10	IM	20.9	14.9	0.9	298	-	[151]
ZIF-68	(bIM)(nIM)	41.3	33.3	0.9	298	1220	[151]
ZIF-69	(cbIM)(nIM)	38.1	25.9	0.9	298	1070	[151]
ZIF-71	dcIM	18.0	19.4	0.9	298	-	[151]
$Cu_2(L)(H_2O)_2$	Pyrazol	32	-	1	195	844.5	[152]
$[Zn_2(L)]$	Pyrazol	37.4	-	1	195	1075.4	[152]
$[Cd_2(L)]$	Pyrazol	24.6	-	1	195	571.7	[152]
$[Co_2(L)(H_2O)_6]$	Pyrazol	31.6	-	1	195	734.6	[152]
Zn4(bpta)2-1	Bipyridine pillar ligands	8.2	34.82	1.2	298	413	[153]
Zn4(bpta)2-1	Bipyridine pillar ligands	3.1	27.69	1.2	298	51	[153]
Cu2L (DMA)4	Acrylamide	22.2	35	1	296	1433	[154]
Zn(ad)(ain)	2-Aminoisonicotinate and adeninate	9.2	40	1	298	399	[155]
bio-MOF-11	Adenine	22.2	33.1	1	273	1148	[156]
bio-MOF-12	Adenine	16.2	38.4	1	273	-	[156]
bio-MOF-13	Adenine	10.4	40.5	1	273	-	[156]
bio-MOF-14	Adenine	8	-	1	273	17	[156]
$Cu(tba)_2$	Triazol	7.3	36	1	293	-	[131]

Table 3. CO_2 uptake for MOFs modified with amine containing ligands

Apart from amine groups, there are other functional moieties that are proven to be effective in enhancing the performance of MOFs in CO_2 capture. Phosphonate and sulfonate organic ligands have gained tremendous attention recently due to their significant improvements in MOF stability toward water [157]. Several studies are reported based on the use of phosphonate and sulfonate ligands, for instance, the selective CO_2/N_2 separation over nitrogen-containing phosphonate MOFs was studied by Marco et al. [100], and the synthesis, stability, porosity of the phosphonate MOFs [158], and their major applications were reported for water stability studies [159–161]. The shielding effect exhibited by phosphonate groups were responsible for the improved stability under humid conditions up to 90% relative humidity at 353 K as observed for CALF-30 [161]. The enhanced water stability of these MOFs was attributed to the kinetic blocking effect which makes the framework completely hydrophobic [159].

MOFs containing nitrogen-donor building blocks were also widely investigated, particularly adenine group which was extensively used due to framework robustness, richness in nitrogen sites, and framework diversity [162]. Song et al. [163] reported the preparation of three new adenine-based MOFs by controlling the adenine coordination with Cd metal sites. This study has provided insight into the controlled synthesis of MOFs by controlling the structure building units (SBU) which can be utilized to extend the idea to include multiple building units within the same framework. Similar studies are also available based on adenine groups as building units, where the effect of the adenine functionalization on framework topology, porosity, and adsorption behavior was investigated [164]. The use of Zn-adeninate SBU led to the discovery of highly porous Bio-MOF-11 to 14 series [165] and Bio-MOF-100 [166] with exceptional surface area (4300 m^2/g) and very large pore volume (4.3cm^3/g); however, the framework stability of these materials still needs to be addressed as the material tends to lose its porosity under harsh activation environment. This issue has been tackled by Zhang et al. [167] to prepare more stable adenine-based PCN-530 structure. Lin et al. have observed high density of open nitrogen-donor sites on 1,3,5-tris(2H-tetrazol-5-yl)benzene (H3BTT) which was responsible for the enhanced CO_2 capacity [146] through the improvement of the framework porosity and the utilization of nitrogen sites readily available to adsorb CO_2. However, the richness of nitrogen atoms in the framework does not necessarily favor CO_2 adsorption, as reported by Gao et al. [110] for the case of tetrazolate-based rth-MOF that has more exposed nitrogen sites as compared to pyrazolate-based rht-MOF and yet was showing less CO_2 uptake attributed to the strong electric field observed on the pyrazolate-based rht-MOF.

Other ligand modifications are also reported in literature by deploying several types of functional groups such as hydroxyl groups (OH) on Zn(BDC) [168], $(CH_3)_2$, (OH), and (COOH) on MIL-53(Al) [145], NO_2 on IRMOF-8 [107] as well as alkyl and nitro groups grafted on DUT-5 [169]. Based on the contribution by Yaghi's group [170], several studies were dedicated to understand the effects of ligand extension on the pore size, surface area, and the sorption behavior of MOFs [98, 109, 133, 171–173]. Recently, zeolite-like MOFs denoted as ZTIFs have attracted great interest due to their unique characteristics for tuning the structure toward various applications [174, 175]. New frameworks (ZTIF-1 and ZTIF-2) were recently reported based on the incorporation of tetrazolates into Zn-Imidazolate structures [176], with similar structures. UTSA-49 was also reported by Chen and coworkers for the selective separation of CO_2/N_2 mixture [117].

Lately, the idea of mixed ligand approach for the synthesis of MOFs with tunable properties has gained much attention which allows for incorporating several functionalities within each ligand to target certain properties such as improving the stability and the capacity for CO_2 simultaneously [177]. For instance, the water stability issue was tackled by Marco and coworkers [178] by utilizing two heterocyclic N-donor-mixed phosphonate-based organic ligands. The designed MOF has shown great water stability and achieved CO_2 uptake of 77 cm^3/g at low pressure and 195 K. By deploying the mixed-ligand approach, Liu et al. [179] have successfully prepared Co-based MOFs containing both benzenetricarboxylic- and triazole-based ligands by using a solvo-thermal synthesis technique. The synthesized MOFs displayed CO_2 uptake of up to 15.2 wt. % at 1 bar and 295 K as well as remarkable selectivity toward nitrogen. A detailed investigation of mixed ligand approach in the design of MOFs is available in literature [180, 181]; however, further work is still needed to optimize the synthesis conditions and correlate the observed performance to the appropriate constituents on the organic ligands. Recent work by Yaghi et al. provided tools to quantitatively map different functional groups incorporated into the same MOF structure [182].

3.2.3. Post-synthetic Functionalization of MOFs

As mentioned previously, tuning the affinity of the framework functionalities toward CO_2 is crucial for improving adsorptive capacities. The aim is to decorate the pore surface in order to have high adsorption selectivity and capacity and yet minimize the regeneration energy. In addition to the pre-synthetic modification of the organic linker, post-synthetic functionalization of MOFs (PSM) is considered a viable route to insert functionalities into the MOF structure after the formation of the basic framework. This approach can overcome the limitations observed in pre-synthetic functionalization, for instance precise control of the synthesis conditions is needed to preserve the functional groups during the solvo-thermal synthesis conditions. Note that some functional groups are not stable under synthesis conditions which require a narrow range of conditions to prepare the MOFs. Others, however, cannot be introduced to the synthesis mixture due to solubility issues, hindrance effects, and they might participate in the crystallization process and yield unwanted materials. Besides, inserting functional groups on the metal sites prior to the synthesis of the framework might intervene in the formation of the building units which can result in the deterioration of the crystal structure [183–185]. PSM is therefore considered an attractive pathway to tailor the properties of MOFs toward better CO_2 capture performance.

In order to make use of high amine affinity toward CO_2, several amine moieties were selected for the modification of various solid sorbents [186–189] including MOFs [190, 191]. Ethylene diamine (en) is considered the most commonly used type of amine for PSM of MOFs for CO_2 capture application. In 2014, Lee et al. [192] reported grafting the diamine into the expanded MOF-74 or Mg(dopbdc) structure at amine loadings of 16.7 wt. % at room temperature which exhibited very high CO_2 uptake of 13.7 wt% at 0.15 bar higher than the 12.1 wt. % capacity reported by McDonald et al. for N,N'-dimethylethylenediamine (mmen) grafted on Mg(dopbdc) [173]. The isosteric heat of adsorption was recorded to be 49 to 51 kJ/mol indicating chemisorption of CO_2 molecules which was further confirmed by the formation of carbamic acid probed by the in situ Fourier transform infrared spectroscopy (FTIR) experiments. The en grafted Mg(dopbdc) was further evaluated for the multicycle adsorption, and it has only lost 3% of its CO_2 uptake after five cycles. Moreover, en-Mg(dopbdc) has also shown stable structure and capacity after exposure to different moisture contents, and therefore this material has a potential for large-scale CO_2 capture (see Figure 5). NH_2, CH_2NH_2, CH_2NHMe along with other functional groups were recently grafted on IRMOF-74, and it was found that IRMOF-74-III-CH_2NH_2 displayed CO_2 capacity of 3.2 mmol/g at 1 bar [132]. The sodalite-type structure Cu-BTTri was also grafted by en functional group [193] which showed chemisorption interaction with the adsorbed CO_2 molecule as can be observed from the high isosteric heat of adsorption (90 kJ/mol). However, the en-Cu-BTTri has only shown improved capacity at low pressure while the unmodified MOF shows higher uptakes at high pressure which is attributed to the significant reduction in Brunauer–Emmett–Teller (BET) surface area from 1770 to 345 m^2/g due to the pore blocking effect of the en group. In an attempt to address this issue, McDonald at al. functionalized mmen group on Cu-BTTri and preserved a BET surface area of 870 m^2/g with 96 kJ/mol isosteric heat of adsorption, nitrogen selectivity of 327, and CO_2 uptake of 9.5 wt.% under 0.15 bar CO_2/0.75 bar N_2 mixture at 25 °C. The negative impact of alkylamine functional groups on reducing the surface area was evident, and one approach to overcome this issue is to introduce ligand extension prior to the introduction of the amine group so as to increase the MOF porosity and avoid the pore-blocking problem during PSMs [132]. Also, a deep insight into the mechanism of CO_2 adsorption on alkylamine-grafted MOFs is crucial to further understand the interactions for improved structural design and amine loadings [194]. Other amine functionalities such as piperazine were also grafted into Cu-BTTri [128] and exhibited 2.5 times higher CO_2 uptake as compared to bare Cu-BTTri, while the heat of adsorption confirms the chemisorption interactions. The area reduction was also evident as it was reduced from 1700 m^2/g to 380 m^2/g (similar to ethylendiamin, (en)- Cu-BTTri [195]). Pyridine was also grafted on Ni-DOBDC to improve the water stability and increase the hydrophobicity of the material [196]. Experimental observations supported by simulations results confirmed the enhanced water stability for the Pyridine-Ni-DOBDC samples while maintaining the CO_2 uptake at atmospheric conditions and low pressures. It was also concluded that the amine moiety was grafted on the unsaturated metal sites of the framework, which makes this approach desirable for amine functionalization. From a combined experimental and simulation study, it was found that pyridine modification of an MOF can reduce H_2O adsorption while retaining considerable CO_2 capacity at conditions of interest for flue gas separation. This indicates that post-synthesis modification of MOFs by coordinating hydrophobic ligands to unsaturated metal sites may be a powerful method to generate new sorbents for gas separation under humid conditions. Amine functionalization to target the water stability of MOFs will be further discussed in the next section.

It is evident from the previous discussion that amine impregnation into MOFs always sacrifices the surface area of the final product. Therefore, the choice of the amine that can improve the affinity toward CO_2 and attain high surface area simultaneously is a trade-off issue. MIL-101 materials were reported to have the highest pore volume and surface area among MOFs to date (BET = 3125 m^2/g and 1.63 cm^3/g). Hence they allow the incorporation of amines with longer alkyl chains such as polyethyleneimine while at the same time maintaining relatively high surface area (1112.6 m^2/g after 75 wt% amine loading). PEI-loaded MIL-101 prepared by Lin et al. [197] exhibited remarkably high CO_2 uptake of 4.2 mmol/g at 0.15 bar and 298 K with exceptional CO_2/N_2 selectivity of 770 at 25 °C.

Optimization of amine loadings and distribution within the MOF structure is a detrimental factor for the impact of these functionalities on the performance in CO_2 capture process. Precise control of the different factors during the grafting process is crucial to append these groups exactly on the unsaturated metal centers, while avoid blocking the pores and hindering access to the interior volume. Improving the PSM methods is considered one of the means to achieve the ideal grafting and amine distribution [191].

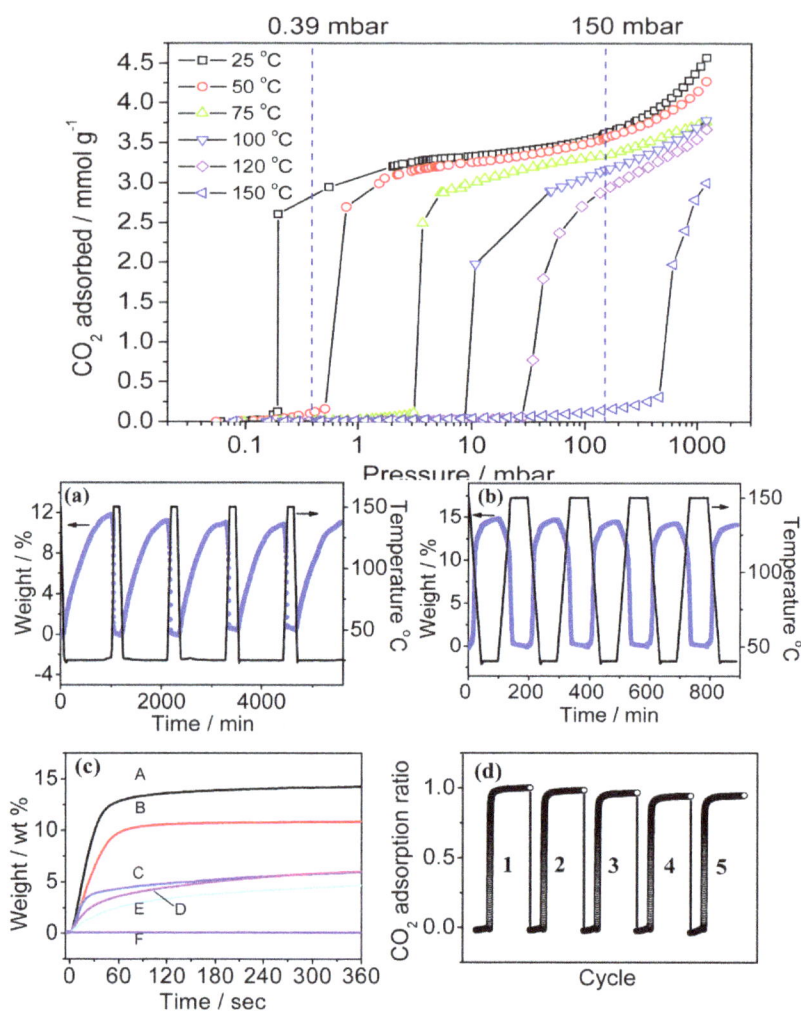

Figure 5. Top: Adsorption isotherms of CO_2 for 1-en at the indicated temperatures. Bottom: Adsorption–desorption cycling of CO_2 for 1-en showing reversible uptake from (a) simulated air (0.39 mbar CO_2 and 21% O_2 balanced with N_2) and from (b) simulated flue gas (0.15 bar CO_2 balanced with N_2). (c) time-dependent CO_2 adsorption for porous materials (A = 1-en, B = mmen-Mg_2(dobpdc), C = 1, D = Mg-MOF-74, E = Zeolite 13X, F = MOF-5). (d) CO_2 adsorption ratio of 1-en in flue gas (after 6 min exposure to 100% RH at 21 °C) to 1-en in flue gas (Adapted from [192]).

4. Recent Advances and Current Trends

4.1. Hybrid Systems Based on MOFs

For more efficient utilization of MOFs sorbents, several hybrid systems based on MOFs with other solid sorbents have been investigated in the literature. The objective of having hybrid materials is to utilize the synergism between the two sorbents and therefore ultimately improve the overall performance in CO_2 separation. Moreover, sorbents such as activated carbons, graphenes, and CNTs provide the added feature of high surface area and easily functionalized sites which contribute to the tuning of the final properties of the composite material. CNTs represent one of the effective candidates that can improve the properties of MOFs for gas adsorption applications. Zhu et al. [198] incorporated HKUST-1 in the interspace of CNTs. The designed composite exhibited superior selectivity and a CO_2 saturation capacity of 7.83 mmol/g at 298 K, which was attributed to the high porosity and surface area. In a similar study, multiwall CNTs, well dispersed in MIL-101 (Cr), were successfully prepared and maintained the same framework and crystal structure as MIL-101. An increase of 60% in CO_2 uptake was observed for the MWCNT-MIL-101 composite which was attributed to the increased porosity as a result of incorporating CNTs [199], as was confirmed by similar work on MWCNT-MIL-53(Cu) composite [200].

Graphene oxide composites with different MOFs are extensively reported in the literature such as HKUST-1 [201], MOF-5 [202], and Cu-BTC [203]. Graphite oxide (GO) is considered a stabilizing agent for MOFs under humid environment, and it has shown remarkable CO_2 capacity of 3.3 mmol/g and great stability under simulated flue gas conditions for GO/Cu-BTC composite [203]. The synthesis of Cu-based MOFs composite with aminated graphite oxides (GO) was carried out and fully characterized by Zhao et al. [204]. The composite exhibited 50% enhanced porosity as compared to the parent MOF and displayed unique structure and pore sizes effective for size exclusion separation of CO_2 from the flue gas. Silica aerogel (SA) was also investigated as a promising candidate for hybrid systems with ZIF-8 [205]. The detailed characterizations of the SA/ZIF-8 confirmed the presence of the two phases in the composite after sol–gel synthesis procedure with different ZIF-8 loadings and mild BET surface area [205].

Several composite materials have been reported for various applications; however, utilizing these hybrid systems in CO_2 adsorption might be a promising route for improving the CO_2 capture process. Ahmed et al. [206] published a review of information related to the synthesis and adsorption applications of MOF composite materials.

4.2. Ionic Liquids/MOF Composites

Ionic liquids as solvents for the absorption separation of CO_2 from flue gas are discussed in Section 1.2 in order to overcome the limitations related to the poor dynamics of CO_2 separation in ILs due to their high viscosity. MOFs can act as an ideal support material for the incorporation of ILs into their porous structure while preserving their unique properties. The concept of immobilization of ILs into solid sorbents has been reported for various applications. For instance, ILs immobilization on mesoporous silica was reported for the catalytic esterification reaction [207], ILs addition into polymer gels for ionic conductivity applications [208], ILs/Zeolite composites [209], in addition to several review papers available on this topic indicating the widespread use of this new approach over the past years [210, 211]. Computational investigation of the theoretical possibility of incorporating ILs into MOFs was studied by Jiang's group for (BMIM)PF6 IL supported on IRMOF-1 for CO_2 capture applications [212]. The confinement effects of the narrow pore on the ILs and the ionic interactions between $[BMIM]^+$ favor the open pore while the anion, $[PF6]^-$, was attached to the open metal sites, was observed in a simulation study. It was ascertained that CO_2 was favorably attached to the $[PF6]^-$ anions sites. The study demonstrated that IL/MOF composites are a potential candidate for CO_2 adsorption and have displayed significantly high CO_2/N_2 selectivity. To the best of our knowledge, the first report on an experimental attempt to immobilize ILs into MOF structures was published by Liu et al. [213] for the insertion of Bronsted acidic ILs (BAIL) into the pores of MIL-101 using post-synthetic approach with triethylene diamine (TEDA) or imidazole (IMIZ) as a solvent assisting during the functionalization process. Nitrogen adsorption isotherms of bare MIL-101, TEDA-BAIL/MIL-101, and all (IMIZ-BAIL/MIL-101) samples showed type-I isotherm indicating the microporous nature of the composite. BET surface area was 1873 m^2/g for the bare MIL-101 which was slightly decreased to 1728 m^2/g and 1148 m^2/g for IMIZ-BAIL/MIL-101 and TEDA-BAIL/MIL-101, respectively. Following this leading report, Jhung's group [214] successfully grafted up to 50 wt.% acidic chloroaluminate IL on MIL-101 which reduced the BET surface area of the bare MIL-101 by around 60%. The incorporation of ILs with basic nature which is favorable for CO_2 adsorption was for the first time reported by Kitagawa et al. [215]. A detailed characterization and investigation of the phase behavior of the immobilized 1-ethyl-3-methylimidazolium bis(trifluoromethylsulfonylamide) denoted as EMI-TFSA ILs into ZIF-8 was presented in this study. A reduction of 29% in pore volume was measured in N_2 adsorption isotherm experiments and computational calculations. The EMI-TFSA/ZIF-8 composite has shown distinctive ion conductivity at low temperature as reported in a second paper by the same group [216]. The prospect of IL/MOF composite for gas separation is still under computational investigation with no reported experimental studies of CO_2 adsorption on these composites. Recently, Vincent-Luna et al. [15] investigated the effects of adding room temperature ILs (RTILs) into the pores of Cu-BTC structures. The adsorption of CO_2, N_2, CH_4, and their mixtures were studied by utilizing various RTILs having the same cation 1-ethyl-3-methylimidazolium [EMIM]$^+$ and different anions such as bis[(trifluoromethyl)-sulfonyl]imide [Tf2N]$^-$, thiocyanate [SCN]$^-$, nitrate [NO3]$^-$, tetrafluoroborate [BF4]$^-$, and hexafluorophosphate [PF6]$^-$. The RTIL/Cu-BTC composite has shown enhanced CO_2 uptakes at low pressures with high CO_2/N_2 selectivity due to the polarization driving force rendering these materials as a promising system for post-combustion CO_2 capture. Another application of IL/MOF composite as a precursor for the preparation of nitrogen and boron–nitrogen (N- and BN-)-decorated porous carbons was recently reported by Aijaz et al. [217] as a novel synthesis strategy.

4.3. Ab/Adsorption in Ionic Liquids/MOF Slurry System

Another approach to utilize the combined synergistic advantages of MOF and IL composites is through a novel hybrid adsorption–absorption technology. This novel technology can provide an efficient approach to utilize the high capacity, selectivity, and low heat of adsorption of the solid sorbents along with the advantages of having a continuous flow process that allows for better heat integration and separation rates in contrast to the conventional batch process used in adsorption-only process. Mass transfer enhancement due to the dispersion of fine solids in liquid solvents was studied and insight into the mechanism and the analysis of different mass transfer resistances were described by Zhang and coworkers [218] which was in agreement with previous findings [219–221]. As far as enhancement of CO_2 capture in slurry systems is concerned, a study dealing with AC particles dispersed in K_2CO_3 aqueous solution was reported by Sumin et al. [222] to investigate the influence of the hydrodynamics on the mass transfer improvements. In a similar work by Rosu et al. [223], AC particles were also

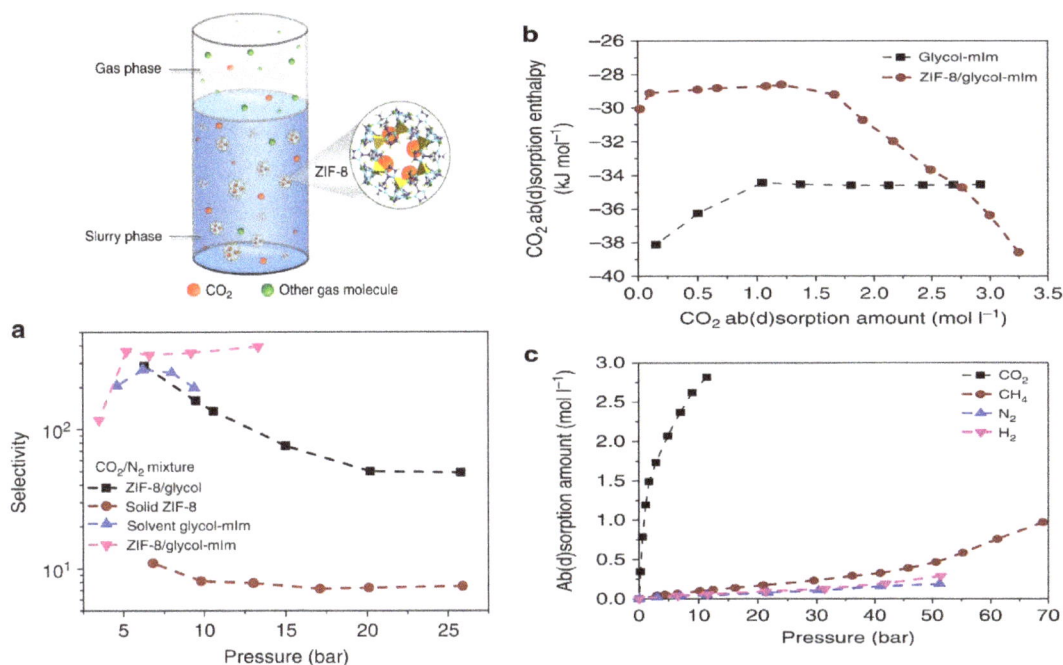

Figure 6. Top left: schematic of the slurry system. (a) Comparison of selectivity toward N_2 (b) ab/adsorption enthalpy. (c) CO_2 uptake at 303.15 K (Adapted from [225]).

found to improve the absorptive CO_2 capture process. The unique characteristics of MOFs in CO_2 adsorption and their recent applications in aqueous solution environment [224] have opened the door toward the possibility of immersing MOF particles in various physical and chemical solvents for CO_2 separation application. This novel unit operation process can overcome the limitations reported for conventional adsorption on MOFs such as high pressure drop and the necessity for formulating the powders into different shapes and sizes which affects their structural stability and reduces the active surface area. Liu et al. [225] reported, for the first time, the preparation of ZIF-8/glycol and ZIF-8/glycol/2-methylimidazole slurries (Figure 6). CO_2 uptake of 1.25 mmol/L was recorded for the slurry system with CO_2/N_2 selectivity of 394 at 1 bar, and most importantly a very low enthalpy of 29 kJ/mol. In a similar work by Lei et al. [226], ILs [EMIM, TF2N] and [OMIM, PF6] were used to prepare slurry systems with ZIF-8 and ZIF-7. CO_2 adsorption in the slurry system has shown a promising performance with isosteric heat of adsorption less than 26 kJ/mol. Following these two studies, the solubility of CO_2 in physical solvents such as methanol mixed with ZIF-8 was also investigated [227]. The study revealed that ZIF-8 can significantly improve the low pressure-CO_2 uptake in physical solvents and can dramatically reduce the solvent losses by evaporation to the gas phase at the top of the absorber. Increasing ZIF-8 loadings has shown further enhancement of the CO_2 capacity, as observed previously [225]; however, it is worth noting that a high solid loading in the slurry system was not recommended from process engineering point of view as it might cause some problems during the pumping of the slurry mixture and increases the solid losses in the multicycle separation process [225].

For future studies on MOF-based slurry systems, there is basic selection of criteria that needs to be satisfied by both MOF and the liquid solution. The selection of the MOF possessing the appropriate pore size for the preparation of the slurry system is very important to guarantee that the size of the liquid is large enough and does not occupy the pores which leaves no space for CO_2 to adsorb. Moreover, the structural stability of the MOF in the aqueous solution is essential so that it does not lose its porous framework nor its surface area. The selection of the liquid candidate is crucial, as it should not provide any extra mass transfer resistance for CO_2 molecules. Further, experimental and computational investigations are still required to understand the separation mechanism in slurry mixtures and to have insight into the different types of interactions between the gas, liquid, and solid materials.

5. Challenges and Outlook

In conclusion, MOFs are considered the largest growing research area in CO_2 capture, with great achievements and developments. Due to their versatile structures and possibilities for various functionalization approaches, the door is still open for further improvements and advancements of their performance under real flue gas conditions, and in large-scale applications. Although we have reported MOFs with distinguished properties and exceptional CO_2 capacity, selectivity, and stability, there are still some concerns that need to be addressed before reaching commercial scale level. The lack of information about the performance of MOFs under real gas mixture conditions is one of the key issues to understand the actual working uptakes and identify any possible limitations. Further experimental testing of MOFs using, for example, a gas mixture containing all the impurities that might be present in an actual flue gas is needed high-throughput technique. Computational gas mixture studies can provide essential information in this regard; however, experimental investigation is still considered the most reliable approach. MOF stabilities in humid conditions, high temperature, and harsh mechanical stress situations must be given much attention. Several studies were performed to target MOF stability, and great achievements were recorded in this field [228, 229], as reviewed in references [160, 230]. Finally, in the following section, we focus on water stability studies as it is one of the main drawbacks of MOFs.

5.1. Water Stability of MOFs

Water stability is a major challenge that has to be overcome before metal organic framework can be used in removing carbon dioxide from flue gas. The core structure of MOF reacts with water vapor content in the flue gas leading to severe distortion of the structure and even failure. As a consequence, the physical structure of MOF is changed, e.g., reduction of porosity and surface area, etc. that decreases the capacity and selectivity for CO_2. Complete dehydration of flue gas increases the cost of separation. It is therefore essential for MOFs to exhibit stability in the presence of water up to certain extent [91].

Metal–ligand coordination bond, which is the most significant part of MOF, is hydrolyzed with water, resulting in the displacement of ligand bond; and as a consequence, the whole structure usually collapses [91]. The stability of MOF in the presence of water depends on the strength of metal ligand bond. pK_a values of the ligand atom can be considered as the strength of this metal- ligand bond. Since the hydrolysis reaction between MOF and water molecule is governed by Gibbs free energy and activation energy of the reactant and product molecules, thermodynamics and kinetics factors have great influence on the water stability of MOF [160]. Insight into the molecular structure, more specifically the metal–ligand strength, the weakest part of MOF, and thermodynamics as well as kinetics study of hydrolysis reaction are very important to improve water stability. Several strategies based on these two important aspects have been taken into consideration.

Jasuja et al. [231] performed a study on the effects of functionalization of the organic ligand in a series of isostructural MOFs in the Zn(BDC-X)-(DABCO)0.5 family on water stability. In this experiment, they cyclically stabilized an unstable parent structure in humid conditions through the incorporation of tetramethyl-BDC ligand. The results of molecular simulation disclosed that the kinetic stability is improved due to the carboxylate oxygen in the DMOF-TM2 structure which acted as a shield to prevent hydrogen-bonding interactions and subsequent structural transformations. Hence, electrophilic zinc atoms in this structure became inaccessible to the nucleophilic oxygen atoms in water, resulting in prevention of the hydrolysis reactions for the displacement ligand. They also performed another study to evaluate the effect of strength of metal–ligand coordination bond and catenation in the framework on water stability [232]. According to their results, the non-interpenetrated MOFs constructed from a pillar ligand of higher pK_a exhibited higher stability; however, interpenetrated MOFs constructed from a pillar ligand of lower pK_a values exhibited less stability. The interpenetration in MOF with incorporation of ligands of relatively high basicity exhibited good water stability. By considering the results of previous experiment, they synthesized cobalt-, nickel-, copper-, and zinc-based, new pillared MOFs of similar topologies which exhibited good water stability [233]. The grafted methyl group on the benzene dicarboxlate (BDC) ligand introduced steric factors around the metal centers; consequently, water stability of MOF drastically improved. The basicity of BTTB-based MOFs synthesized with bipyridyl pillar ligands had lower basicity than DABCO; however, they exhibited better stability in the presence of humid condition.

Bae et al. [114] performed a study to modify Ni-DOBDC with pyridine molecules. The study showed that pyridine molecule made the normally hydrophilic internal surface more hydrophobic; as a result, water absorption was reduced, while substantial CO_2 capture capacity was retained to a certain level. Fracaroli et al. [132] improved the interior of IRMOF-74-III by covalently functionalizing it with a primary amine, and used a MOF, IRMOF-74-IIICH2NH2, for the selective capture of CO_2 in 65% relative humidity.

Zhang et al. [234] performed a study to modify the surface of the MOF hydrophilic to hydrophobic to improve water stability. They demonstrated a new strategy to modify hydrophobic polydimethysiloxane (PDMS) on the surface to significantly enhance their water resistance by a facile vapor deposition technique. In this study, they successfully coated three vulnerable MOFs according to the water stability (MOF-5, HKUST-1, and ZnBT), while the porosity, crystalline characteristics, and surface area were unchanged.

All these studies demonstrated that water stability of MOFs can be improved by incorporating specific factors (e.g., metal–ligand strength, thermodynamic and kinetic factors, etc.) which govern the structural stability of the framework.

Author details

Mohanned Mohamedali, Devjyoti Nath, Hussameldin Ibrahim and Amr Henni*

*Address all correspondence to: amr.henni@uregina.ca

Industrial/Process Systems Engineering, Faculty of Engineering and Applied Science, University of Regina, Regina, SK, Canada

References

[1] World Meteorological Organization, *WMO Greenhouse Gas Bulletin No. 10*, Geneva, Switzerland 2014.

[2] Energy Information Administration (EIA), *Annual Energy Outlook*, Washington, United States, 2014.

[3] A. A. Olajire, *Energy*, 2010, 35, 2610–2628.

[4] J. D. Figueroa, T. Fout, S. Plasynski, H. McIlvried and R. D. Srivastava, *International Journal of Greenhouse Gas Control*, 2008, 2, 9–20.

[5] M. Wang, A. Lawal, P. Stephenson, J. Sidders and C. Ramshaw, *Chemical Engineering Research and Design*, 2011, 89, 1609–1624.

[6] G. Puxty, R. Rowland, A. Allport, Q. Yang, M. Bown, R. Burns, M. Maeder and M. Attalla, *Environmental Science & Technology*, 2009, 43, 6427–6433.

[7] C.-H. Yu, C.-H. Huang and C.-S. Tan, *Aerosol and Air Quality Research*, 2012, 12, 745–769.

[8] R. Idem, M. Wilson, P. Tontiwachwuthikul, A. Chakma, A. Veawab, A. Aroonwilas and D. Gelowitz, *Industrial & Engineering Chemistry Research*, 2006, 45, 2414–2420.

[9] G. T. Rochelle, *Science*, 2009, 325, 1652–1654.

[10] S. Rackley, *Carbon Capture and Storage*, Gulf Professional Publishing, Houston, USA 2009.

[11] Q. Zhuang, R. Pomalis, L. Zheng and B. Clements, *Energy Procedia*, 2011, 4, 1459–1470.

[12] Z. Lei, C. Dai and B. Chen, *Chemical Reviews*, 2013, 114, 1289–1326.

[13] M. Ramdin, A. Amplianitis, S. Bazhenov, A. Volkov, V. Volkov, T. J. Vlugt and T. W. de Loos, *Industrial & Engineering Chemistry Research*, 2014, 53, 15427–15435.

[14] J. F. Brennecke and B. E. Gurkan, *The Journal of Physical Chemistry Letters*, 2010, 1, 3459–3464.

[15] J. M. Vicent-Luna, J. J. Gutiérrez-Sevillano, J. A. Anta and S. Calero, *The Journal of Physical Chemistry C*, 2013, 117, 20762–20768.

[16] L. Zhou, J. Fan and X. Shang, *Materials*, 2014, 7, 3867–-3880.

[17] R. D. Rogers and K. R. Seddon, *Science*, 2003, 302, 792–793.

[18] M. Ramdin, T. W. de Loos and T. J. Vlugt, *Industrial & Engineering Chemistry Research*, 2012, 51, 8149–8177.

[19] E. D. Bates, R. D. Mayton, I. Ntai and J. H. Davis, *Journal of the American Chemical Society*, 2002, 124, 926–927.

[20] Z.-Z. Yang, Y.-N. Zhao and L.-N. He, *RSC Advances*, 2011, 1, 545–567.

[21] H. G. Joglekar, I. Rahman and B. D. Kulkarni, *Chemical Engineering & Technology*, 2007, 30, 819–828.

[22] J.-R. Li, Y. Ma, M. C. McCarthy, J. Sculley, J. Yu, H.-K. Jeong, P. B. Balbuena and H.-C. Zhou, *Coordination Chemistry Reviews*, 2011, 255, 1791–1823.

[23] A. W. Chester and E. G. Derouane, *Zeolite characterization and catalysis*, Springer, 2009.

[24] A. Primo and H. Garcia, *Chemical Society Reviews*, 2014, 43, 7548–7561.

[25] T. F. Degnan Jr, *Topics in Catalysis*, 2000, 13, 349–356.

[26] W. Hölderich, J. Röseler, G. Heitmann and A. Liebens, *Catalysis Today*, 1997, 37, 353–366.

[27] M. Davis, *Microporous and Mesoporous Materials*, 1998, 21, 173–182.

[28] S. Cavenati, C. A. Grande and A. E. Rodrigues, *Energy & Fuels*, 2006, 20, 2648–2659.

[29] S. Himeno, T. Tomita, K. Suzuki and S. Yoshida, *Microporous and Mesoporous Materials*, 2007, 98, 62–69.

[30] R. V. Siriwardane, M.-S. Shen, E. P. Fisher and J. A. Poston, *Energy & Fuels*, 2001, 15, 279–284.

[31] R. V. Siriwardane, M.-S. Shen and E. P. Fisher, *Energy & Fuels*, 2003, 17, 571–576.

[32] J. Merel, M. Clausse and F. Meunier, *Industrial & Engineering Chemistry Research*, 2008, 47, 209–215.

[33] T. Remy, S. A. Peter, L. Van Tendeloo, S. Van der Perre, Y. Lorgouilloux, C. E. Kirsch-hock, G. V. Baron and J. F. Denayer, *Langmuir*, 2013, 29, 4998–5012.

[34] J. Shang, G. Li, R. Singh, P. Xiao, J. Z. Liu and P. A. Webley, *The Journal of Physical Chemistry C*, 2013, 117, 12841–12847.

[35] F. Su and C. Lu, *Energy & Environmental Science*, 2012, 5, 9021–9027.

[36] S.-H. Hong, M.-S. Jang, S. J. Cho and W.-S. Ahn, *Chemical Communications*, 2014, 50, 4927–4930.

[37] O. Cheung, Z. Bacsik, Q. Liu, A. Mace and N. Hedin, *Applied Energy*, 2013, 112, 1326–1336.

[38] A. Goj, D. S. Sholl, E. D. Akten and D. Kohen, *The Journal of Physical Chemistry B*, 2002, 106, 8367–8375.

[39] S. Loganathan, M. Tikmani and A. K. Ghoshal, *Langmuir*, 2013, 29, 3491–3499.

[40] C. F. Cogswell, H. Jiang, J. Ramberger, D. Accetta, R. J. Willey and S. Choi, *Langmuir*, 2015, 31, 4534–4541.

[41] R. Veneman, H. Kamphuis and D. Brilman, *Energy Procedia*, 2013, 37, 2100–2108.

[42] D. P. Bezerra, R. S. Oliveira, R. S. Vieira, C. L. Cavalcante Jr and D. C. Azevedo, *Adsorption*, 2011, 17, 235–246.

[43] S. C. Lee, C. C. Hsieh, C. H. Chen and Y. S. Chen, *Aerosol and Air Quality Research*, 2013, 13, 360–366.

[44] Y. Jing, L. Wei, Y. Wang and Y. Yu, *Microporous and Mesoporous Materials*, 2014, 183, 124–133.

[45] Y.-K. Kim, Y.-H. Mo, J. Lee, H.-S. You, C.-K. Yi, Y. C. Park and S.-E. Park, *Journal of Nanoscience and Nanotechnology*, 2013, 13, 2703–2707.

[46] K. Ahmad, O. Mowla, E. M. Kennedy, B. Z. Dlugogorski, J. C. Mackie and M. Stock-enhuber, *Energy Technology*, 2013, 1, 345–349.

[47] C. H. Lee, D. H. Hyeon, H. Jung, W. Chung, D. H. Jo, D. K. Shin and S. H. Kim, *Journal of Industrial and Engineering Chemistry*, 2015, 23, 251–256.

[48] J. Kim, L.-C. Lin, J. A. Swisher, M. Haranczyk and B. Smit, *Journal of the American Chemical Society*, 2012, 134, 18940–18943.

[49] D. Marx, L. Joss, M. Hefti, R. Pini and M. Mazzotti, *Energy Procedia*, 2013, 37, 107–114.

[50] G. Li, P. Xiao, P. Webley, J. Zhang, R. Singh and M. Marshall, *Adsorption*, 2008, 14, 415-422.

[51] A. Sayari and Y. Belmabkhout, *Journal of the American Chemical Society*, 2010, 132, 6312–6314.

[52] S. Esposito, A. Marocco, G. Dell'Agli, B. De Gennaro and M. Pansini, *Microporous and Mesoporous Materials*, 2015, 202, 36–43.

[53] R. A. Fiuza Jr, R. Medeiros de Jesus Neto, L. B. Correia and H. M. Carvalho Andrade, *Journal of Environmental Management*, 2015, 161, 198–205.

[54] E. David and J. Kopac, *Journal of Analytical and Applied Pyrolysis*, 2014, 110, 322–332.

[55] F. Montagnaro, A. Silvestre-Albero, J. Silvestre-Albero, F. Rodríguez-Reinoso, A. Erto, A. Lancia and M. Balsamo, *Microporous and Mesoporous Materials*, 2015, 209, 157–164.

[56] M. Kacem, M. Pellerano and A. Delebarre, *Fuel Processing Technology*, 2015, 138, 271–283.

[57] D. Xu, P. Xiao, J. Zhang, G. Li, G. Xiao, P. A. Webley and Y. Zhai, *Chemical Engineering Journal*, 2013, 230, 64–72.

[58] G. Sethia and A. Sayari, *Carbon*, 2015, 93, 68–80.

[59] N. Díez, P. Álvarez, M. Granda, C. Blanco, R. Santamaría and R. Menéndez, *Chemical Engineering Journal*, 2015, 281, 704–712.

[60] R.-L. Tseng, F.-C. Wu and R.-S. Juang, *Separation and Purification Technology*, 2015, 140, 53–60.

[61] A. Houshmand, M. S. Shafeeyan, A. Arami-Niya and W. M. A. W. Daud, *Journal of the Taiwan Institute of Chemical Engineers*, 2013, 44, 774–779.

[62] M. J. Mostazo-López, R. Ruiz-Rosas, E. Morallón and D. Cazorla-Amorós, *Carbon*, 2015, 91, 252–265.

[63] J. A. A. Gibson, A. V. Gromov, S. Brandani and E. E. B. Campbell, *Microporous and Mesoporous Materials*, 2015, 208, 129–139.

[64] M. S. Shafeeyan, W. M. A. W. Daud, A. Shamiri and N. Aghamohammadi, *Chemical Engineering Research and Design*, 2015, 104, 42–52.

[65] Q. Liu, Y. Shi, S. Zheng, L. Ning, Q. Ye, M. Tao and Y. He, *Journal of Energy Chemistry*, 2014, 23, 111–118.

[66] M.-S. Lee and S.-J. Park, *Journal of Solid State Chemistry*, 2015, 226, 17–23.

[67] M.-S. Lee, S.-Y. Lee and S.-J. Park, *International Journal of Hydrogen Energy*, 2015, 40, 3415–3421.

[68] F. Su, C. Lu, W. Cnen, H. Bai and J. F. Hwang, *Science of The Total Environment*, 2009, 407, 3017–3023.

[69] A. Kumar Mishra and S. Ramaprabhu, *RSC Advances*, 2012, 2, 1746–1750.

[70] M. M. Gui, Y. X. Yap, S.-P. Chai and A. R. Mohamed, *International Journal of Greenhouse Gas Control*, 2013, 14, 65–73.

[71] Y. G. Ko, H. J. Lee, H. C. Oh and U. S. Choi, *Journal of Hazardous Materials*, 2013, 250–251, 53–60.

[72] K. S. Novoselov, A. K. Geim, S. V. Morozov, D. Jiang, Y. Zhang, S. V. Dubonos, I. V. Grigorieva and A. A. Firsov, *Science*, 2004, 306, 666–669.

[73] S.-M. Hong, S. H. Kim and K. B. Lee, *Energy & Fuels*, 2013, 27, 3358–3363.

[74] M. Asai, T. Ohba, T. Iwanaga, H. Kanoh, M. Endo, J. Campos-Delgado, M. Terrones, K. Nakai and K. Kaneko, *Journal of the American Chemical Society*, 2011, 133, 14880–14883.

[75] L.-Y. Meng and S.-J. Park, *Journal of Colloid and Interface Science*, 2012, 386, 285–290.

[76] F. Li, X. Jiang, J. Zhao and S. Zhang, *Nano Energy*, 2015, 16, 488–515.

[77] S. Gadipelli and Z. X. Guo, *Progress in Materials Science*, 2015, 69, 1–60.

[78] A. Taheri Najafabadi, *Renewable and Sustainable Energy Reviews*, 2015, 41, 1515–1545.

[79] K. C. Kemp, V. Chandra, M. Saleh and K. S. Kim, *Nanotechnology*, 2013, 24, 235703.

[80] J. Oh, Y.-H. Mo, V.-D. Le, S. Lee, J. Han, G. Park, Y.-H. Kim, S.-E. Park and S. Park, *Carbon*, 2014, 79, 450–456.

[81] W. Li, X. Jiang, H. Yang and Q. Liu, *Applied Surface Science*, 2015, 356, 812–816.

[82] S. Chowdhury, G. K. Parshetti and R. Balasubramanian, *Chemical Engineering Journal*, 2015, 263, 374–384.

[83] J. Wang, X. Mei, L. Huang, Q. Zheng, Y. Qiao, K. Zang, S. Mao, R. Yang, Z. Zhang, Y. Gao, Z. Guo, Z. Huang and Q. Wang, *Journal of Energy Chemistry*, 2015, 24, 127–137.

[84] Y. Zhao, H. Ding and Q. Zhong, *Applied Surface Science*, 2013, 284, 138–144.

[85] Y. Cao, Y. Zhao, Z. Lv, F. Song and Q. Zhong, *Journal of Industrial and Engineering Chemistry*, 2015, 27, 102–107.

[86] S. Pourebrahimi, M. Kazemeini, E. Ganji Babakhani and A. Taheri, *Microporous and Mesoporous Materials*, 2015, 218, 144–152.

[87] J.-R. Li, R. J. Kuppler and H.-C. Zhou, *Chemical Society Reviews*, 2009, 38, 1477–1504.

[88] T. Duren, Y.-S. Bae and R. Q. Snurr, *Chemical Society Reviews*, 2009, 38, 1237–1247.

[89] S. Kitagawa, R. Kitaura and S.-i. Noro, *Angewandte Chemie International Edition*, 2004, 43, 2334–2375.

[90] O. M. Yaghi, M. O'Keeffe, N. W. Ockwig, H. K. Chae, M. Eddaoudi and J. Kim, *Nature*, 2003, 423, 705–714.

[91]   K. Sumida, D. L. Rogow, J. A. Mason, T. M. McDonald, E. D. Bloch, Z. R. Herm, T.-H. Bae and J. R. Long, *Chemical reviews*, 2011, 112, 724–781.

[92]   K. C. Stylianou and W. L. Queen, *CHIMIA International Journal for Chemistry*, 2015, 69, 274–283.

[93]   Z. Zhang, Z.-Z. Yao, S. Xiang and B. Chen, *Energy & Environmental Science*, 2014, 7, 2868–2899.

[94]   J. Wang, L. Huang, R. Yang, Z. Zhang, J. Wu, Y. Gao, Q. Wang, D. O'Hare and Z. Zhong, *Energy & Environmental Science*, 2014, 7, 3478–3518.

[95]   J.-R. Li, J. Sculley and H.-C. Zhou, *Chemical Reviews*, 2011, 112, 869–932.

[96]   J.-R. Li, R. J. Kuppler and H.-C. Zhou, *Chemical Society Reviews*, 2009, 38, 1477–1504.

[97]   L. Li, S. Tang, C. Wang, X. Lv, M. Jiang, H. Wu and X. Zhao, *Chemical Communications*, 2014, 50, 2304–2307.

[98]   J. Cai, X. Rao, Y. He, J. Yu, C. Wu, W. Zhou, T. Yildirim, B. Chen and G. Qian, *Chemical Communications*, 2014, 50, 1552–1554.

[99]   M. Taddei, F. Costantino, F. Marmottini, A. Comotti, P. Sozzani and R. Vivani, *Chemical Communications*, 2014, 50, 14831–14834.

[100] M. Taddei, F. Costantino, A. Ienco, A. Comotti, P. V. Dau and S. M. Cohen, *Chemical Communications*, 2013, 49, 1315–1317.

[101] Y. Peng, G. Srinivas, C. E. Wilmer, I. Eryazici, R. Q. Snurr, J. T. Hupp, T. Yildirim and O. K. Farha, *Chemical Communications*, 2013, 49, 2992–2994.

[102] J. Kim, Y.-R. Lee and W.-S. Ahn, *Chemical Communications*, 2013, 49, 7647–7649.

[103] Y. He, H. Furukawa, C. Wu, M. O'Keeffe, R. Krishna and B. Chen, *Chemical Communications*, 2013, 49, 6773–6775.

[104] X. Wu, M. N. Shahrak, B. Yuan and S. Deng, *Microporous and Mesoporous Materials*, 2014, 190, 189–196.

[105] G. Yang, J. A. Santana, M. E. Rivera-Ramos, O. García-Ricard, J. J. Saavedra-Arias, Y. Ishikawa, A. J. Hernández-Maldonado and R. G. Raptis, *Microporous and Mesoporous Materials*, 2014, 183, 62–68.

[106] E. Deniz, F. Karadas, H. A. Patel, S. Aparicio, C. T. Yavuz and M. Atilhan, *Microporous and Mesoporous Materials*, 2013, 175, 34–42.

[107] S. Orefuwa, E. Iriowen, H. Yang, B. Wakefield and A. Goudy, *Microporous and Mesoporous Materials*, 2013, 177, 82–90.

[108] S. Ye, X. Jiang, L.-W. Ruan, B. Liu, Y.-M. Wang, J.-F. Zhu and L.-G. Qiu, *Microporous and Mesoporous Materials*, 2013, 179, 191–197.

[109] N. C. Burtch, H. Jasuja, D. Dubbeldam and K. S. Walton, *Journal of the American Chemical Society*, 2013, 135, 7172–7180.

[110] W.-Y. Gao, T. Pham, K. A. Forrest, B. Space, L. Wojtas, Y.-S. Chen and S. Ma, *Chemical Communications*, 2015, 51, 9636–9639.

[111] D. Wang, B. Liu, S. Yao, T. Wang, G. Li, Q. Huo and Y. Liu, *Chemical Communications*, 2015.

[112] S. K. Elsaidi, M. H. Mohamed, H. T. Schaef, A. Kumar, M. Lusi, T. Pham, K. A. Forrest, B. Space, W. Xu and G. J. Halder, *Chemical Communications*, 2015.

[113] Y.-W. Li, J. Xu, D.-C. Li, J.-M. Dou, H. Yan, T.-L. Hu and X.-H. Bu, *Chemical Communications*, 2015, 51, 14211–14214.

[114] Y.-S. Bae, J. Liu, C. E. Wilmer, H. Sun, A. N. Dickey, M. B. Kim, A. I. Benin, R. R. Willis, D. Barpaga and M. D. LeVan, *Chemical Communications*, 2014, 50, 3296–3298.

[115] K. Liu, B. Li, Y. Li, X. Li, F. Yang, G. Zeng, Y. Peng, Z. Zhang, G. Li and Z. Shi, *Chemical Communications*, 2014, 50, 5031–5033.

[116] R. Li, X. Ren, X. Feng, X. Li, C. Hu and B. Wang, *Chemical Communications*, 2014, 50, 6894–6897.

[117] F. Wang, H.-R. Fu, Y. Kang and J. Zhang, *Chemical Communications*, 2014, 50, 12065–12068.

[118] S. Xiong, Y. Gong, H. Wang, H. Wang, Q. Liu, M. Gu, X. Wang, B. Chen and Z. Wang, *Chemical Communications*, 2014, 50, 12101–12104.

[119] C. Song, Y. He, B. Li, Y. Ling, H. Wang, Y. Feng, R. Krishna and B. Chen, *Chemical Communications*, 2014, 50, 12105–12108.

[120] J. Qian, F. Jiang, K. Su, J. Pan, Z. Xue, L. Liang, P. P. Bag and M. Hong, *Chemical Communications*, 2014, 50, 15224–15227.

[121] C. H. Lau, R. Babarao and M. R. Hill, *Chemical Communications*, 2013, 49, 3634–3636.

[122] J. Luo, J. Wang, G. Li, Q. Huo and Y. Liu, *Chemical Communications*, 2013, 49, 11433–11435.

[123] J. A. Johnson, Q. Lin, L.-C. Wu, N. Obaidi, Z. L. Olson, T. C. Reeson, Y.-S. Chen and J. Zhang, *Chemical Communications*, 2013, 49, 2828–2830.

[124] V. Bon, J. Pallmann, E. Eisbein, H. C. Hoffmann, I. Senkovska, I. Schwedler, A. Schneemann, S. Henke, D. Wallacher and R. A. Fischer, *Microporous and Mesoporous Materials*, 2015.

[125] S. Chaemchuen, K. Zhou, N. A. Kabir, Y. Chen, X. Ke, G. Van Tendeloo and F. Verpoort, *Microporous and Mesoporous Materials*, 2015, 201, 277–285.

[126] E. Keceli, M. Hemgesberg, R. Grünker, V. Bon, C. Wilhelm, T. Philippi, R. Schoch, Y. Sun, M. Bauer and S. Ernst, *Microporous and Mesoporous Materials*, 2014, 194, 115–125.

[127] L. Li, J. Yang, J. Li, Y. Chen and J. Li, *Microporous and Mesoporous Materials*, 2014, 198, 236–246.

[128] A. Das, M. Choucair, P. D. Southon, J. A. Mason, M. Zhao, C. J. Kepert, A. T. Harris and D. M. D'Alessandro, *Microporous and Mesoporous Materials*, 2013, 174, 74–80.

[129] R. Sabouni, H. Kazemian and S. Rohani, *Microporous and Mesoporous Materials*, 2013, 175, 85–91.

[130] X. Wu, Z. Bao, B. Yuan, J. Wang, Y. Sun, H. Luo and S. Deng, *Microporous and Mesoporous Materials*, 2013, 180, 114–122.

[131] M. Du, C.-P. Li, M. Chen, Z.-W. Ge, X. Wang, L. Wang and C.-S. Liu, *Journal of the American Chemical Society*, 2014, 136, 10906–10909.

[132] A. M. Fracaroli, H. Furukawa, M. Suzuki, M. Dodd, S. Okajima, F. Gándara, J. A. Reimer and O. M. Yaghi, *Journal of the American Chemical Society*, 2014, 136, 8863–8866.

[133] X. Zhao, X. Bu, Q.-G. Zhai, H. Tran and P. Feng, *Journal of the American Chemical Society*, 2015, 137, 1396–1399.

[134] W. Cai, T. Lee, M. Lee, W. Cho, D.-Y. Han, N. Choi, A. C. Yip and J. Choi, *Journal of the American Chemical Society*, 2014, 136, 7961–7971.

[135] J. A. Mason, T. M. McDonald, T.-H. Bae, J. E. Bachman, K. Sumida, J. J. Dutton, S. S. Kaye and J. R. Long, *Journal of the American Chemical Society*, 2015, 137, 4787–4803.

[136] A. L. Dzubak, L.-C. Lin, J. Kim, J. A. Swisher, R. Poloni, S. N. Maximoff, B. Smit and L. Gagliardi, *Nat Chem*, 2012, 4, 810–816.

[137] X. Kong, E. Scott, W. Ding, J. A. Mason, J. R. Long and J. A. Reimer, *Journal of the American Chemical Society*, 2012, 134, 14341–14344.

[138] A. O. Yazaydın, R. Q. Snurr, T.-H. Park, K. Koh, J. Liu, M. D. LeVan, A. I. Benin, P. Jakubczak, M. Lanuza and D. B. Galloway, *Journal of the American Chemical Society*, 2009, 131, 18198–18199.

[139] P. Canepa, C. A. Arter, E. M. Conwill, D. H. Johnson, B. A. Shoemaker, K. Z. Soliman and T. Thonhauser, *Journal of Materials Chemistry A*, 2013, 1, 13597–13604.

[140] X.-J. Hou, P. He, H. Li and X. Wang, *The Journal of Physical Chemistry C*, 2013, 117, 2824–2834.

[141] C. R. Wade and M. Dincă, *Dalton Transactions*, 2012, 41, 7931–7938.

[142] P. L. Llewellyn, S. Bourrelly, C. Serre, A. Vimont, M. Daturi, L. Hamon, G. De Weireld, J.-S. Chang, D.-Y. Hong and Y. Kyu Hwang, *Langmuir*, 2008, 24, 7245–7250.

[143] C. P. Cabello, P. Rumori and G. T. Palomino, *Microporous and Mesoporous Materials*, 2014, 190, 234–239.

[144] L. Li, J. Yang, J. Li, Y. Chen and J. Li, *Microporous and Mesoporous Materials*, 2014, 198, 236–246.

[145] A. Torrisi, R. G. Bell and C. Mellot-Draznieks, *Microporous and Mesoporous Materials*, 2013, 168, 225–238.

[146] Q. Lin, T. Wu, S.-T. Zheng, X. Bu and P. Feng, *Journal of the American Chemical Society*, 2011, 134, 784–787.

[147] E. Keceli, M. Hemgesberg, R. Grünker, V. Bon, C. Wilhelm, T. Philippi, R. Schoch, Y. Sun, M. Bauer, S. Ernst, S. Kaskel and W. R. Thiel, *Microporous and Mesoporous Materials*, 2014, 194, 115–125.

[148] Y. Yang, R. Lin, L. Ge, L. Hou, P. Bernhardt, T. E. Rufford, S. Wang, V. Rudolph, Y. Wang and Z. Zhu, *Dalton Transactions*, 2015, 44, 8190–8197.

[149] R. Vaidhyanathan, S. S. Iremonger, K. W. Dawson and G. K. Shimizu, *Chemical communications*, 2009, 5230–5232.

[150] W.-Y. Gao, W. Yan, R. Cai, L. Meng, A. Salas, X.-S. Wang, L. Wojtas, X. Shi and S. Ma, *Inorganic chemistry*, 2012, 51, 4423–4425.

[151] L. Ding and A. O. Yazaydin, *Physical Chemistry Chemical Physics*, 2013, 15, 11856–11861.

[152] C. J. Coghlan, C. J. Sumby and C. J. Doonan, *CrystEngComm*, 2014, 16, 6364–6371.

[153] Z.-H. Xuan, D.-S. Zhang, Z. Chang, T.-L. Hu and X.-H. Bu, *Inorganic Chemistry*, 2014, 53, 8985–8990.

[154] R.-R. Cheng, S.-X. Shao, H.-H. Wu, Y.-F. Niu, J. Han and X.-L. Zhao, *Inorganic Chemistry Communications*, 2014, 46, 226–228.

[155] E. Yang, H.-Y. Li, F. Wang, H. Yang and J. Zhang, *CrystEngComm*, 2013, 15, 658–661.

[156] T. Li, D.-L. Chen, J. E. Sullivan, M. T. Kozlowski, J. K. Johnson and N. L. Rosi, *Chemical Science*, 2013, 4, 1746–1755.

[157] G. K. Shimizu, R. Vaidhyanathan and J. M. Taylor, *Chemical Society Reviews*, 2009, 38, 1430–1449.

[158] R. K. Mah, B. S. Gelfand, J. M. Taylor and G. K. Shimizu, *Inorganic Chemistry Frontiers*, 2015, 2, 273–277.

[159] J. M. Taylor, R. Vaidhyanathan, S. S. Iremonger and G. K. Shimizu, *Journal of the American Chemical Society*, 2012, 134, 14338–14340.

[160] N. C. Burtch, H. Jasuja and K. S. Walton, *Chemical Reviews*, 2014, 114, 10575–10612.

[161] B. S. Gelfand, J.-B. Lin and G. K. Shimizu, *Inorganic Chemistry*, 2015, 54, 1185–1187.

[162] S. Verma, A. K. Mishra and J. Kumar, *Accounts of Chemical Research*, 2009, 43, 79–91.

[163] Y. Song, X. Yin, B. Tu, Q. Pang, H. Li, X. Ren, B. Wang and Q. Li, *CrystEngComm*, 2014, 16, 3082–3085.

[164] J. Thomas-Gipson, G. Beobide, O. Castillo, M. Froᐩba, F. Hoffmann, A. Luque, S. Pérez-Yáñez and P. Román, *Crystal Growth & Design*, 2014, 14, 4019–4029.

[165] T. Li, D.-L. Chen, J. E. Sullivan, M. T. Kozlowski, J. K. Johnson and N. L. Rosi, *Chemical Science*, 2013, 4, 1746–1755.

[166] J. An, O. K. Farha, J. T. Hupp, E. Pohl, J. I. Yeh and N. L. Rosi, *Nature Communications*, 2012, 3, 604.

[167] M. Zhang, W. Lu, J.-R. Li, M. Bosch, Y.-P. Chen, T.-F. Liu, Y. Liu and H.-C. Zhou, *Inorganic Chemistry Frontiers*, 2014, 1, 159–162.

[168] Y. Zhao, H. Wu, T. J. Emge, Q. Gong, N. Nijem, Y. J. Chabal, L. Kong, D. C. Langreth, H. Liu and H. Zeng, *Chemistry-A European Journal*, 2011, 17, 5101–5109.

[169] M. A. Gotthardt, S. Grosjean, T. S. Brunner, J. Kotzel, A. M. Ganzler, S. Wolf, S. Brase and W. Kleist, *Dalton Transactions*, 2015, 44, 16802–16809.

[170] J. Kim, B. Chen, T. M. Reineke, H. Li, M. Eddaoudi, D. B. Moler, M. O'Keeffe and O. M. Yaghi, *Journal of the American Chemical Society*, 2001, 123, 8239–8247.

[171] L. Du, Z. Lu, K. Zheng, J. Wang, X. Zheng, Y. Pan, X. You and J. Bai, *Journal of the American Chemical Society*, 2012, 135, 562–565.

[172] D. Wang, B. Liu, S. Yao, T. Wang, G. Li, Q. Huo and Y. Liu, *Chemical Communications*, 2015, 51, 15287–15289.

[173] T. M. McDonald, W. R. Lee, J. A. Mason, B. M. Wiers, C. S. Hong and J. R. Long, *Journal of the American Chemical Society*, 2012, 134, 7056–7065.

[174] M. Eddaoudi, D. F. Sava, J. F. Eubank, K. Adil and V. Guillerm, *Chemical Society Reviews*, 2015, 44, 228–249.

[175] B. A. Al-Maythalony, O. Shekhah, R. Swaidan, Y. Belmabkhout, I. Pinnau and M. Eddaoudi, *Journal of the American Chemical Society*, 2015, 137, 1754–1757.

[176] F. Wang, H.-R. Fu, Y. Kang and J. Zhang, *Chemical Communications*, 2014, 50, 12065–12068.

[177] Z. Zhang, Y. Zhao, Q. Gong, Z. Li and J. Li, *Chemical Communications*, 2013, 49, 653–661.

[178] M. Taddei, F. Costantino, A. Ienco, A. Comotti, P. V. Dau and S. M. Cohen, *Chemical Communications*, 2013, 49, 1315–1317.

[179] B. Liu, R. Zhao, K. Yue, J. Shi, Y. Yu and Y. Wang, *Dalton Transactions*, 2013, 42, 13990–13996.

[180] X.-L. Zhao and W.-Y. Sun, *CrystEngComm*, 2014, 16, 3247–3258.

[181] M. Du, C.-P. Li, C.-S. Liu and S.-M. Fang, *Coordination Chemistry Reviews*, 2013, 257, 1282–1305.

[182] X. Kong, H. Deng, F. Yan, J. Kim, J. A. Swisher, B. Smit, O. M. Yaghi and J. A. Reimer, *Science*, 2013, 341, 882–885.

[183] J. Liu, P. K. Thallapally, B. P. McGrail, D. R. Brown and J. Liu, *Chemical Society Reviews*, 2012, 41, 2308–2322.

[184] Z. Wang, K. K. Tanabe and S. M. Cohen, *Inorganic Chemistry*, 2008, 48, 296–306.

[185] H. Deng, C. J. Doonan, H. Furukawa, R. B. Ferreira, J. Towne, C. B. Knobler, B. Wang and O. M. Yaghi, *Science*, 2010, 327, 846–850.

[186] T. Watabe, Y. Nishizaka, S. Kazama and K. Yogo, *Energy Procedia*, 2013, 37, 199–204.

[187] T. Watabe and K. Yogo, *Separation and Purification Technology*, 2013, 120, 20–23.

[188] T. Prenzel, M. Wilhelm and K. Rezwan, *Chemical Engineering Journal*, 2014, 235, 198–206.

[189] M. Auta and B. H. Hameed, *Chemical Engineering Journal*, 2014, 253, 350–355.

[190] R. K. Motkuri, J. Liu, C. A. Fernandez, S. K. Nune, P. Thallapally and B. P. McGrail, *Industrial Catalysis and Separations*, 2014, 61–103.

[191] S. M. Cohen, *Chemical Reviews*, 2011, 112, 970–1000.

[192] W. R. Lee, S. Y. Hwang, D. W. Ryu, K. S. Lim, S. S. Han, D. Moon, J. Choi and C. S. Hong, *Energy & Environmental Science*, 2014, 7, 744–751.

[193] A. Demessence, D. M. D'Alessandro, M. L. Foo and J. R. Long, *Journal of the American Chemical Society*, 2009, 131, 8784–8786.

[194] N. Planas, A. L. Dzubak, R. Poloni, L.-C. Lin, A. McManus, T. M. McDonald, J. B. Neaton, J. R. Long, B. Smit and L. Gagliardi, *Journal of the American Chemical Society*, 2013, 135, 7402–7405.

[195] T. M. McDonald, D. M. D'Alessandro, R. Krishna and J. R. Long, *Chemical Science*, 2011, 2, 2022–2028.

[196] Y.-S. Bae, J. Liu, C. E. Wilmer, H. Sun, A. N. Dickey, M. B. Kim, A. I. Benin, R. R. Willis, D. Barpaga, M. D. LeVan and R. Q. Snurr, *Chemical Communications*, 2014, 50, 3296–3298.

[197] Y. Lin, Q. Yan, C. Kong and L. Chen, *Scientific Reports*, 2013, 3.

[198] L. Ge, L. Wang, V. Rudolph and Z. Zhu, *RSC Advances*, 2013, 3, 25360–25366.

[199] M. Anbia and V. Hoseini, *Chemical Engineering Journal*, 2012, 191, 326–330.

[200] M. Anbia and S. Sheykhi, *Journal of Industrial and Engineering Chemistry*, 2013, 19, 1583–1586.

[201] X. Zhou, W. Huang, J. Miao, Q. Xia, Z. Zhang, H. Wang and Z. Li, *Chemical Engineering Journal*, 2015, 266, 339–344.

[202] C. Petit and T. J. Bandosz, *Advanced Materials*, 2009, 21, 4753–4757.

[203] Z. Bian, J. Xu, S. Zhang, X. Zhu, H. Liu and J. Hu, *Langmuir*, 2015, 31, 7410–7417.

[204] Y. Zhao, M. Seredych, Q. Zhong and T. J. Bandosz, *RSC Advances*, 2013, 3, 9932–9941.

[205] A. Prabhu, A. Al Shoaibi and C. Srinivasakannan, *Materials Letters*, 2015, 146, 43–46.

[206] I. Ahmed and S. H. Jhung, *Materials Today*, 2014, 17, 136–146.

[207] B. Zhen, Q. Jiao, Y. Zhang, Q. Wu and H. Li, *Applied Catalysis A: General*, 2012, 445–446, 239–245.

[208] W. Lu, K. Henry, C. Turchi and J. Pellegrino, *Journal of The Electrochemical Society*, 2008, 155, A361–A367.

[209] S. Ntais, A. Moschovi, V. Dracopoulos and V. Nikolakis, *ECS Transactions*, 2010, 33, 41–47.

[210] H. Li, P. S. Bhadury, B. Song and S. Yang, *RSC Advances*, 2012, 2, 12525–12551.

[211] B. Xin and J. Hao, *Chemical Society Reviews*, 2014, 43, 7171–7187.

[212] Y. Chen, Z. Hu, K. M. Gupta and J. Jiang, *The Journal of Physical Chemistry C*, 2011, 115, 21736–21742.

[213] Q.-X. Luo, M. Ji, M.-H. Lu, C. Hao, J.-S. Qiu and Y.-Q. Li, *Journal of Materials Chemistry A*, 2013, 1, 6530–6534.

[214] N. A. Khan, Z. Hasan and S. H. Jhung, *Chemistry-A European Journal*, 2014, 20, 376–-380.

[215] K. Fujie, T. Yamada, R. Ikeda and H. Kitagawa, *Angewandte Chemie International Edition*, 2014, 53, 11302–11305.

[216] K. Fujie, K. Otsubo, R. Ikeda, T. Yamada and H. Kitagawa, *Chemical Science*, 2015.

[217] A. Aijaz, T. Akita, H. Yang and Q. Xu, *Chemical Communications*, 2014, 50, 6498–6501.

[218] G. D. Zhang, W. F. Cai, C. J. Xu and M. Zhou, *Chemical Engineering Science*, 2006, 61, 558–568.

[219] O. Ozkan, A. Calimli, R. Berber and H. Oguz, *Chemical Engineering Science*, 2000, 55, 2737–2740.

[220] H. Vinke, P. J. Hamersma and J. M. H. Fortuin, *Chemical Engineering Science*, 1993, 48, 2197–2210.

[221] A. A. C. M. Beenackers and W. P. M. Van Swaaij, *Chemical Engineering Science*, 1993, 48, 3109–3139.

[222] S. Lu, Y. Ma, C. Zhu and S. Shen, *Chinese Journal of Chemical Engineering*, 2007, 15, 842–846.

[223] M. Rosu, A. Marlina, A. Kaya and A. Schumpe, *Chemical Engineering Science*, 2007, 62, 7336–7343.

[224] B. Chen, Y. Ji, M. Xue, F. R. Fronczek, E. J. Hurtado, J. U. Mondal, C. Liang and S. Dai, *Inorganic Chemistry*, 2008, 47, 5543–5545.

[225] H. Liu, B. Liu, L.-C. Lin, G. Chen, Y. Wu, J. Wang, X. Gao, Y. Lv, Y. Pan and X. Zhang, *Nature Communications*, 2014, 5.

[226] Z. Lei, C. Dai and W. Song, *Chemical Engineering Science*, 2015, 127, 260–268.

[227] C. Dai, W. Wei and Z. Lei, *Journal of Chemical & Engineering Data*, 2015, 60, 1311–1317.

[228] B. Van de Voorde, I. Stassen, B. Bueken, F. Vermoortele, D. De Vos, R. Ameloot, J.-C. Tan and T. D. Bennett, *Journal of Materials Chemistry A*, 2015, 3, 1737–1742.

[229] H. Wu, T. Yildirim and W. Zhou, *The Journal of Physical Chemistry Letters*, 2013, 4, 925–930.

[230] M. Bosch, M. Zhang and H.-C. Zhou, *Advances in Chemistry*, 2014, 2014.

[231] H. Jasuja, N. C. Burtch, Y.-g. Huang, Y. Cai and K. S. Walton, *Langmuir*, 2013, 29, 633–642.

[232] H. Jasuja and K. S. Walton, *Dalton Transactions*, 2013, 42, 15421–15426.

[233] H. Jasuja, Y. Jiao, N. C. Burtch, Y.-G. Huang and K. S. Walton, *Langmuir*, 2014, 30, 14300–14307.

[234] W. Zhang, Y. Hu, J. Ge, H.-L. Jiang and S.-H. Yu, *Journal of the American Chemical Society*, 2014, 136, 16978–16981.

Mitigating Greenhouse Gas Emissions from Winter Production of Agricultural Greenhouses

Lilong Chai, Chengwei Ma, Baoju Wang, Mingchi Liu and Zhanhui Wu

Additional information is available at the end of the chapter

Abstract

Consuming conventional fossil fuel, such as coal, natural gas, and oil, to heat agricultural greenhouses has contributed to the climate change and air pollutions regionally and globally, so the clean energy sources have been increasingly applied to replace fossil energies in heating agricultural greenhouses, especially in urban area. To assess the environment performance (e.g., greenhouse gas (GHG) emissions) of the ground source heat pump system (GSHPs) for heating agricultural greenhouses in urban area, a GSHPs using the shallow geothermal energy (SGE) in groundwater was applied to heat a Chinese solar greenhouse (G1) and a multispan greenhouse (G2) in Beijing (latitude 39°40′ N), the capital city of China. Emission rates of the GSHPs for heating the G1 and G2 were quantified to be 0.257–0.879 g CO_2 eq. m^{-2} day^{-1}. The total GHG emissions from heating greenhouses in Beijing with the GSHPs were quantified as 1.7–2.9 Gt CO_2 eq. $year^{-1}$ based on the electricity from the coal-fired power plant (CFPP) and the gas-fired power plant (GFPP). Among different stages of the SGE flow, the SGE promotion contributed most GHG emissions (66%) in total due to the higher consumption of electricity in compressors. The total GHG emissions from greenhouses heating with the coal-fired heating system (CFHs) and gas-fired heating system (GFHs) were quantified as 2.3–5.2 Gt CO_2 eq. $year^{-1}$ in Beijing. Heating the G1 and G2 with the GSHPs powered by the electricity from the CFPP, the equivalent CO_2 emissions were 43% and 44% lower than directly burning coal with the CFHs but were 46% and 44% higher than the GFHs that burn natural gas. However, when using the GFPP-generated electricity to run the GSHPs, the equivalent CO_2 emissions would be 84% and 47% lower than the CFHs and the GFHs, respectively.

Keywords: urban agriculture, greenhouse heating, greenhouse gases, fossil energy, shallow geothermal energy

1. Introduction

Agricultural buildings, such as horticultural greenhouse, usually require additional heating during winter and cold days in high latitude regions of the Northern Hemisphere [1, 2]. In northern China and many European countries, coal-fired heating system (CFHs) and the natural gas-fired heating system (GFHs) are dominant heating methods in greenhouses [3, 4]. However, conventional fossil fuels, such as coal, natural gas, and oil, which are nonrenewable and are the major greenhouse gas (GHG) contributors, may lead to the global climate change, air pollution, and energy crisis [5-7].

Renewable and clean energy, such as solar, geothermal, and shallow geothermal energy (SGE), has been increasingly applied to replace fossil energy systems in heating agricultural buildings (especially in urban area) across the world [8-11]. The SGE is mainly the stored solar energy in groundwater and soil layers less than 200 m deep from the earth soil surface [12, 13]. It can be used as heat source or sink for air conditioning in residential, industrial, and agricultural buildings with the ground source heat pump system (GSHPs), also known as geothermal heat pumps (GHPs) [4, 14].

The GSHPs has been applied to heat agricultural greenhouse in many countries [15-18]. The GSHPs could be considered with zero GHG emissions if the electricity was the only energy source that could be consumed by the system. However, producing electricity in coal-fired power plant (CFPP) or gas-fired power plant (GFPP) would emit a large quantity of GHG (e.g., CO_2). Besides, the refrigerant (e.g., R22 and R134a) used by the heat pump unit has been reported with the high risk of leaking in the year-round operation [19]. Therefore, assessing GHG emissions of the GSHPs should consider both direct and indirect sources.

In northern China (the area with altitude higher than 30° in the Northern Hemisphere) [20], there was mainly two kinds of horticultural greenhouses: the Chinese solar greenhouse (denoted as G1), which may or may not require assisted heating depending on the building design and the plants be cultivated, and the multispan greenhouses (denoted as G2), which require 100% assisted heating systems (primarily in the form of coal burning or gas burning) during winter time [21, 22]. The Chinese solar greenhouse, characterized with east-west orientation, transparent camber south roof, and solid north roof and east and west walls, usually has higher heat-preserving capacity than multispan greenhouse and requires less heating [23, 24]. However, the healthy growth of thermophilic vegetables, such as cucumber and tomato, and most flowers in Chinese solar greenhouse still requires assisted heating especially during cold winter nights or consecutive days of snowing or cloudy [3].

By the end of 2007, about 19,300 ha of greenhouses and tunnels had been constructed and used in Beijing, the capital city of China [25, 26], primarily for producing vegetables, flowers, and fruits. About 6000 ha Chinese solar greenhouses (the structure similar to G1) and 1000 ha multispan greenhouses (the structure similar to G2) may require assisted heating in winter with the systems of the CFHs and GFHs [27]. Therefore, quantifying the heating rate and GHG emission rate for the primary types of agricultural greenhouses with different heating systems and energy sources is important for developing the national or regional GHG emissions inventory of the greenhouse heating and mitigation strategies.

The objectives of this chapter are to (1) address the environmental concern on agricultural production over winter; (2) quantify the heating loads and the GHG emission rates for the two primary agricultural greenhouses (the G1-Chinese solar greenhouse and the G2-multispan greenhouses) in northern China; (3) assess the annual GHG emissions inventory of the greenhouse heating with different energy sources in Beijing, the capital city of the China; and (4) identify the difference between the shallow geothermal energy and the conventional fossil energy systems in GHG emissions of agricultural greenhouses heating.

2. Materials and Methods

2.1. Selected/tested greenhouses and the GSHPs

A Chinese solar greenhouse (G1, Figure 1 and Table 1) and a multispan greenhouse (G2, Figure 2 and Table 1), two important types of greenhouse in Northern China, were equipped with the groundwater-type GSHPs (Figure 3 and Table 2) in Beijing (latitude 39°40′N) and tested for developing heating rate and GHG emission rate. Performances of GHSPs were compared to CFHs and GFHs. In addition, different electricity generation methods (e.g., coal and gas power plant) were considered for assessing the GHG emissions of the GSHPs.

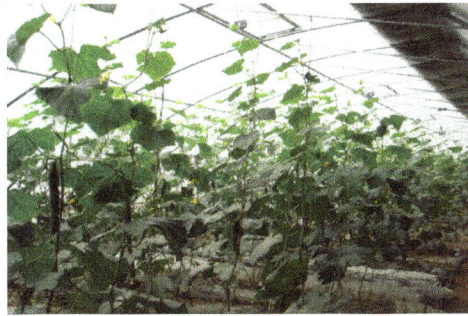

(a) Cucumbers in the G1

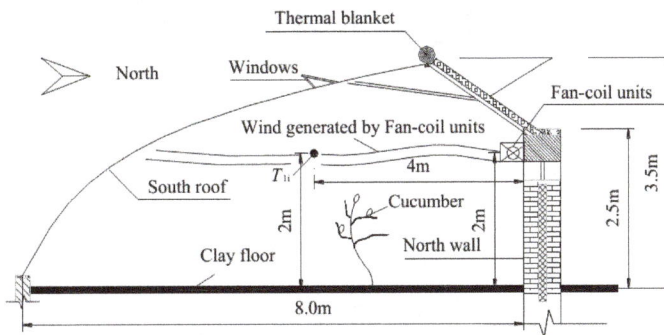

(b) The profile of the G1

Figure 1. The Chinese solar greenhouse (G1)

Section	Construction and coverage material	Surface area, m²
Chinese solar greenhouse (G1)		
(1) North wall	240 mm clay brick+100 mm polystyrene heat preservation layer+240 mm clay brick	150
(2) East and west end walls	240 mm clay brick+100 mm polystyrene heat preservation foam board+240 mm clay brick	36
(3) South roof	0.15 mm single layer transparent polyethylene (south roof was covered with 10 mm needled felt heat blanket at winter nights)	510
(4) North roof	50 mm steel plate+100 polystyrene heat preservation layer+50 mm steel plate	108
(5) Floor	Bare soil (clay)	480
Multispan greenhouses (G2)		
(1) North wall	5 mm coated steel sheet+100 mm polystyrene heat preservation layer+5 mm coated steel sheet	126
(2) East and west wall	20 mm double-layer glass	110
(3) South wall	20 mm double-layer glass	98
(4) North roof	20 mm double-layer glass	108
(5) Floor	Bare soil for planting with concrete walkway (floor area was covered with aluminum film at horizontal height of 3.5 m at winter nights)	756

Table 1. Characteristics of the testing greenhouses.

(a) Cottons in the G2

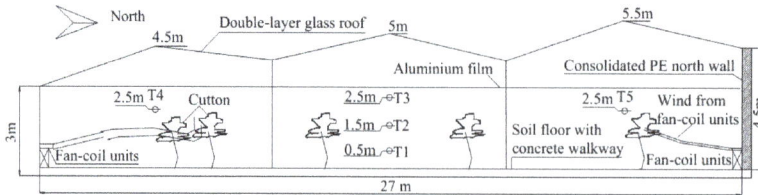

(b) The profile of the G2

Figure 2. The multispan greenhouse.

(a) The heat pump and water circulating system

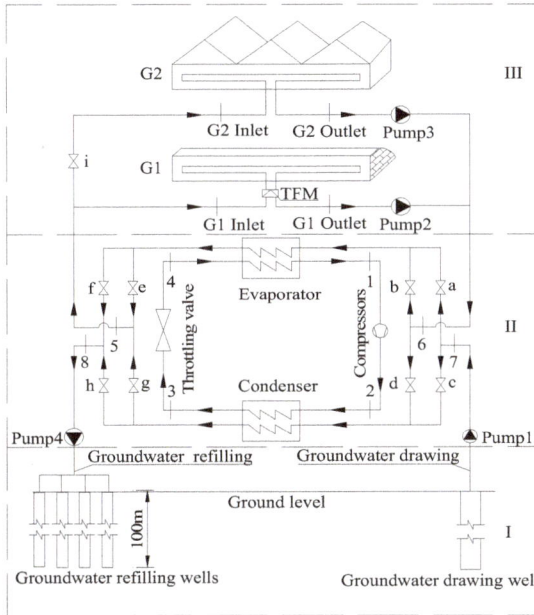

(b) Diagram of the GSHPs greenhouse heating system

Figure 3. The groundwater type of GSHPs. (I is the stage of SGE extraction, II is the stage of SGE promotion, and III is the stage of greenhouse heating. **a–i** are valves installed in different water pipes. Pump 1 is groundwater drawing pump; pumps 2 and 3 are circulating water pumps; and pump 4 represent the groundwater backfilling pump. TFM is the thermal flow meter installed position. 1–8 represent the thermodynamic points in each section of GSHPs [18].

The GSHPs started to heat greenhouses on October 15, 2007 and ended on February 4, 2008. Cucumbers and strawberries were grown in G1 and cotton was grown in G2 during heating tests. The fan coil units in two greenhouses were controlled by the T-type thermocouple controllers automatically. For G1, the indoor air temperature was controlled in the range of 18°C–20°C, and for G2, the indoor air temperature was controlled in the range of 18°C–22°C, considering the poor thermal stability caused by the larger volume in G2.

Elements	Performance parameters
(1) Compressor	Manufacturer: four Danfoss hermetic scroll compressor; rated power of electric motor driving: 16.08×4 kW; refrigerant: 58 kg R22. Rate of refrigeration capacity: 380 kW; rate of heating capacity: 450 kW.
(2) Condenser	Horizontal shell–tube model
(3) Evaporator	Dry evaporator model
(4) Throttle	Copper capillary tube
(5) Fan-coil	In G1: FP-136; rated input power: 56 W; number: 6. In G2: 42VM006; rated input power: 87 W; number: 45
(6) Pumps	Flux: 33.2 m^3 h^{-1}; rated input power: 11 kW; number: 4
(7) Control system	PLC touching screen controller

Table 2. Characteristics of the GSHPs.

2.2. Power inputs and heating rate quantification

The heat provided by the GSHPs for heating greenhouses was quantified with the thermal flow meter (TFM) (Model DN35 and DN100, Beijing Jingyuan Liquid Apparatus Company, Beijing, China). A weather station (Qingsheng Electronic Science and Technology Co. Ltd., Handan, China) installed in the agriculture station was used to monitor outdoor environmental factors. Total electricity consumption of the GSHPs was recorded with watt-hour meter (Shanghai Huaxia Ammeters Manufactory, Shanghai, China).

2.3. Quantifying the GHG emissions

There are six kinds of gases (Table 3) highlighted as the GHG in Intergovernmental Panel on Climate Change (IPCC 2006) [5]. For the greenhouse heating with the CFHs and GFHs, carbon dioxide (CO_2) is the only GHG to be considered. But for the GSHPs greenhouse heating, most electricity consumed (in Beijing area) was generated in the CFPP or the GFPP, and the process of the power generation could emit large amount of CO_2 [25]. Besides, it has been reported that the leaking fraction of the refrigerant (i.e., R22(HFC-22) in this study) used in the GSHPs is around 0.02 kg^{-1} (2%) per year [19]. The R22 was not listed as one of the six primary GHG in the IPCC (2006), but it was reported with 1.28 times of the global warming potential (GWP) of the R134a (HFC-134a) [28]. Therefore, the GWP of R22 was estimated to be 4902, 1830, and 557 based on 20a, 100a, and 500a, respectively, based on the relationship of the GWP with the R134a.

The GHG emissions from heating G1 and G2 with the GSHPs can be quantified with Eq. 1.

$$\text{EM}_{\text{GSHPs-Gi-}j} = M_{\text{GSHPs-Gi-}j} * f_{j-\text{co}_2} + \frac{HE_{\text{GSHPs,Gi}}}{HE_{\text{GSHPs,G}}} M_{\text{R22,Gi}} * f_{\text{R22-leak}} * f_{\text{R22-co}_2} \tag{1}$$

where $HE_{\text{GSHPs,G}}$, total energy provided by the GSHPs for heating G1 and G2 during whole winter (monitored with TFM), MJ; $HE_{\text{GSHPs,Gi}}$, total energy provided by the GSHPs for heating

GHG	20a	100a	500a
CO_2	1	1	1
CH_4	72	25	7.6
N_2O	289	298	153
HFCs(HFC-134a)	3830	1430	435
PFCs (PFC-116)	8630	22,800	32,600
SF6	16,300	22,800	32,600

Note: Data were cited from IPCC 2006[5], expressed in CO_2 eq.

Table 3. GWP of different greenhouse gases

the greenhouse of type i (G1 or G2) during whole winter (monitored with TFM), MJ; $EM_{GSHPs-Gi-j}$, equivalent CO_2 emissions from the GSHPs for heating greenhouse i (G1 or G2) be driven by electricity generated in power plant j (CFPP or GFPP), kg CO_2 eq.; $M_{GSHPS-Gi-j}$, the coal or natural gas consumed in power plant j (CFPP or GFPP) for generating the electricity that the GSHPs had used for heating greenhouse i (G1 or G2), kg C or kg CH_4; $M_{R22,Gi}$, the amount of the refrigerant R22 in the GSHPs be allotted for heating greenhouse i (G1 or G2), kg; f_{j-} co_{22}, the CO_2 emissions coefficient of fossil energy (coal or natural gas) in power plant j (CFPP or GFPP), kg CO_2 eq. (kg C)$^{-1}$ or kg CO_2 eq. (kg CH_4)$^{-1}$; f_{R22-} co_2 the equivalent CO_2 emissions coefficient of R22, kg CO_2 eq. kg^{-1} R22; and $f_{R22-leak}$, the leaking fraction of total R22, %.

The carbon (C) or natural gas (assumed to be 100% as CH_4 in calculating CO_2 emissions) consumed to produce the electricity consumed by the GSHPs was estimated for the CFPP and GFPP based on Eq. 2.

$$M_{GSHPs-Gi-j} = \frac{3600 \ ELE_{GSHPs,Gi}}{CV_j} \qquad (2)$$

where $ELE_{GSHPs,Gi}$, total electricity consumed by the GSHPs in heating greenhouse i (G1 or G2) during winter production, kWh; and CV_j, conversion factor between heat and electricity in CFPP or GFPP, 0.27 was used for CFPP, and 0.42 was applied to GFPP in this study [29].

The GHG emissions from the GSHPs were compared with two primary greenhouse heating systems used in northern China: the CFHs and the CFHs. The equivalent quantity of CO_2 emissions from the GSHPs, CFHs, and GFHs was quantified based on the same heat energy provided by the GSHPs during whole winter for the G1 and G2 (Eqs. 3–6). In northern China, the heating efficiencies of CFHs and GFHs were considered as 0.6 and 0.8, respectively [27, 29].

$$M_{carbon-Gi} = \frac{HE_{GSHPs,Gi}}{F_{carbon-heat} * f_{C-HE}} \qquad (3)$$

$$M_{\text{Gas}-Gi} = \frac{HE_{\text{GSHPs},Gi}}{F_{\text{gas}-\text{heat}} * f_{G-HE}} \tag{4}$$

$$EM_{\text{CFHs},Gi} = M_{\text{Carbon}-Gi} * f_{C-\text{co}_2} \tag{5}$$

$$EM_{\text{GFHs},Gi} = M_{\text{Gas}-Gi} * f_{G-\text{co}_2} \tag{6}$$

where EM_{CFHs}, CO_2 emissions from greenhouse CFHs, kg CO_2; EM_{GFHs}, CO_2 emissions from greenhouse GFHs, kg CO_2; $f_{C-\text{CO}2}$, CO_2 emissions coefficient of the CFHs, kg CO_2 (kg C)$^{-1}$; $f_{G-\text{co}2}$, CO_2 emissions coefficient of the natural GFHs, kg CO_2 (kg CH_4)$^{-1}$; f_{C-HE}, the heating efficiency of the CFHs, dimensionless; f_{G-HE}, the heating efficiency of the GFHs, dimensionless; $F_{\text{carbon}-\text{heat}}$, specific calorific value of burning per kg carbon, MJ (kg C)$^{-1}$; $F_{\text{carbon}-\text{heat}}$, specific calorific value of burning per kg natural gas, MJ (kg CH_4)$^{-1}$; $M_{\text{carbon}-Gi}$, the carbon consumed in heating greenhouse i (G1 or G2) during whole winter, kg C; $M_{\text{gas}-Gi}$, the natural gas consumed in heating greenhouse i (G1 or G2) during whole winter, kg CH_4; and $M_{\text{R22}-Gi}$, the R22 used in the GSHPs be attributed to greenhouse i (G1 or G2) based on the proportion of heat energy received by G1 and G2, kg.

Under the normal temperature and atmospheric pressure (288 K and 1 atm), burning a kilogram of the standard coal (C) and natural gas (CH_4) in oxygen (O_2) completely has potential to emit 3.67 and 2.75 kg CO_2, respectively, with Eqs. 7 and 8 [30]. Meanwhile, the $F_{\text{carbon}-\text{heat}}$ and $F_{\text{gas}-\text{heat}}$ were quantified as 29.3 and 52.6 MJ for burning each kilogram of the C and CH_4, respectively.

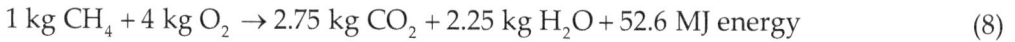

$$1\,\text{kg C} + 2.67\,\text{kg O}_2 \rightarrow 3.67\,\text{kg CO}_2 + 29.3\,\text{MJ energy} \tag{7}$$

$$1\,\text{kg CH}_4 + 4\,\text{kg O}_2 \rightarrow 2.75\,\text{kg CO}_2 + 2.25\,\text{kg H}_2\text{O} + 52.6\,\text{MJ energy} \tag{8}$$

Total GHG emissions from greenhouses heating in Beijing can be estimated based on the total area of greenhouses (similar to G1 or G2) and heat rate per square greenhouse floor (Eqs. 9–11):

$$EMG_{\text{Beijing,GSHPs},j} = \frac{A_{\text{G1, Beijing}} \times f_{\text{G1,heating}} EM_{\text{GSHPs,G1}-\text{elej}}}{480} + \frac{A_{\text{G2, Beijing}} \times f_{\text{G2,heating}} EM_{\text{GSHPs,G2}-\text{elej}}}{756} \tag{9}$$

$$EMG_{\text{Beijing,Carbon}} = \frac{A_{\text{G1, Beijing}} \times f_{\text{G2,heating}} EM_{\text{CFHs,G1}}}{480} + \frac{A_{\text{G2, Beijing}} \times f_{\text{G2,heating}} EM_{\text{CFHs,G2}}}{756} \tag{10}$$

$$EMG_{Beijing,gas} = \frac{A_{G1, Beijing} \times f_{G2,heating} EM_{GFHs,G1}}{480} + \frac{A_{G2, Beijing} \times f_{G2,heating} EM_{GFHs,G2}}{756} \qquad (11)$$

Where, $EMG_{Beijing,GSHPs, j}$, equivalent CO_2 emissions from the GSHPs heating for greenhouses similar to G1 and G2 be driven electricity generated in the power plant j (CFPP or GFPP), kg CO_2 eq. in Beijing; $EMG_{Beijing,carbon}$, equivalent CO_2 emissions from the CFHs heating for greenhouses similar to G1 and G2 in Beijing, kg CO_2 eq; and $EMG_{Beijing,gas}$, equivalent CO_2 emissions from the GFHs heating for greenhouses similar to G1 and G2 in Beijing, kg CO_2 eq.

3. Results and Discussion

3.1. Emissions from the GSHPs

Total heat energy provided by the GSHPs for G1 and G2 were 149270.4 and 640659.1 MJ during 2007–2008 winter at the electricity consumptions of 10826.1 and 44372.2 kWh, respectively. The electricity consumed by the GSHPs usually came from the power plants of CFPP and GFPP in northern China. Therefore, the difference between the CFPP and GFPP in producing GHG at different stages of the SGE in the GSHPs was compared (Table 4).

	SGE extraction	SGE promotion	SGE heating	Inventory
Electricity from CFPP				
G1 electricity consumption, kWh	2440.6	7135.5	1250.3	10826.4
G1 standard coal consumption, kg C	1095.5	3202.7	561.2	4859.4
G1 GHG emissions, kg CO_2 eq.	4020.3	11754.0	2059.5	17833.8
G2 electricity consumption, kWh	10003.0	29245.0	5124.2	44372.2
G2 standard coal consumption, kg C	4489.8	13126.4	2300.0	19916.2
G2 GHG emission, kg CO_2 eq.	16477.4	48174.0	8440.9	73092.3
Electricity from GFPP				
G1 electricity consumption, kWh	2440.6	7135.5	1250.3	10826.4
G1 natural gas consumption, kg CH_4	397.7	1162.8	203.7	1764.2
G1 GHG emissions, kg CO_2 eq.	1093.7	3197.6	560.3	4851.6
G2 electricity consumption, kWh	10003.0	29245.0	5124.2	44372.2
G2 natural gas consumption, kg CH_4	1630.0	4765.6	835.0	7230.7
G2 GHG emission, kg CO_2 eq.	4482.6	13105.4	2296.3	19884.3

Note: (1) G1=480m², heated with GSHPs for 146 days; G2=756 m², heated with GSHPs for 111 days. (2) In calculating the CO_2 emissions, the CH_4 was assumed to be 100% chemical component of the natural gas.

Table 4. GHG emission from G1 and G2 heating with GSHPs.

In producing the amount of the electricity consumed by the GSHPs for G1 and G2, about 4.9 and 20 t of coal (C) were consumed in CFPP. If use the GFPP produced electricity, the total natural gas burned could be 1.8 and 7.2 t (CH_4). During 2007–2008 winter heating, the GWP of the GHG emissions from G1 and G2 (Figure 4) was estimated to be 18.3 and 74.9 t CO_2 eq. with the CFPP and 5.3 and 21.7 t CO_2 eq. with the GFPP, respectively, over a 20-year time horizon. The GWP of 100a was 1.5%–1.6% lower than 20a for G1 and 5.3%–5.4 % lower than 20a for G2 due to the reduced GWP on R22. Similar to HCFC-22, R22 has shorter atmospheric lifetime [5]. Generally, the CO_2 eq. contributed by the leak of R22 accounted for 2.4% and 8.4% in the scenarios of CFPP and GFPP, respectively.

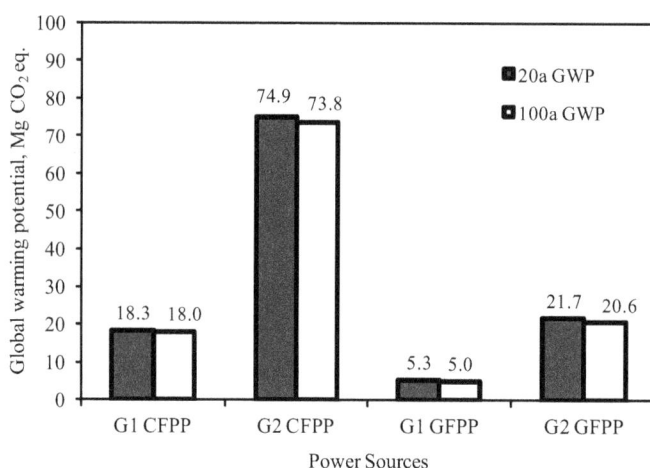

Figure 4. GWP of the GHG emissions derived from the GHSPs heating in G1 and G2.

Among different stages of the SGE flow, most GHG emissions (66%) happened at the stage of SGE promotion due to the higher consumption of electricity in compressors. Therefore, improving the efficiency of the compressors has the potential to reduce the GHG emissions from the GHSPs heating.

3.2. Greenhouse gas emissions from fossil energy systems

Providing G1 and G2 with the same quantity of heat that the GSHPs has provided (i.e., 149270.4 and 640659.1 MJ) requires the CFHs to consume 8.49 and 36.40 t of standard coal and the GFHs to consume 3.55 mg (4964 m^3 at 288 K and 1 atm) and 15.22 t (21,304 m^3 at 288 K and 1 atm) of natural gas (CH_4), respectively. Accordingly, the GHG emissions from heating G1 and G2 (Figure 5) were estimated to be 32.7 and 133.7 t CO_2 eq. for the CFHs system and 9.8 and 41.8 t CO_2 eq. for the GFHs system.

3.3. Standardized GHG emissions from greenhouse heating

The unit electricity consuming rate of the GSHPs was 0.15 and 0.53 kWh m^{-2} d^{-1} for heating G1 and G2 and which can be standardized as 1500 and 5300 kWh ha^{-1} d^{-1} in Beijing during

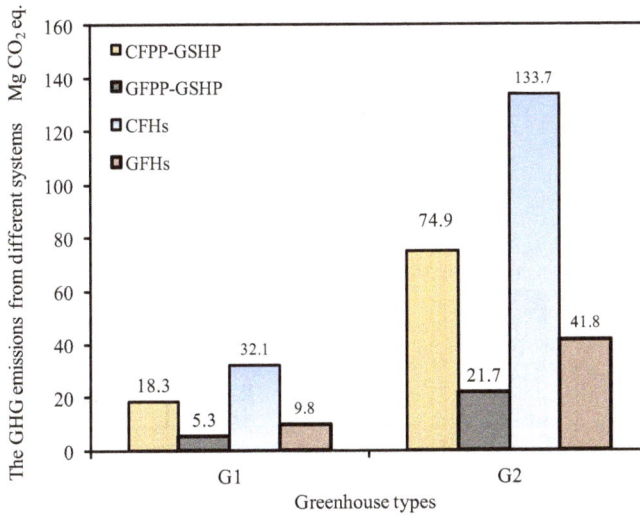

Figure 5. The GHG emissions from the GHSPs and the fossil energy systems.

2007–2008 winter. The 20a and 100a GWPs of the GHG emissions from the GSHPs heating for G1 and G2 (Figure 6a) were 0.076–0.893 and 0.072–0.879 g CO_2 eq. m^{-2} d^{-1} respectively. The GHG emission rate of G2 is 3.42 times of G1 because Chinese solar greenhouse has better heat-preserving capacity than multispan greenhouses.

Regarding to the CFHs and GFHs, the standardized GWPs of the GHG emissions (Figure 6b) were 0.142–1.214 CO_2 eq. m^{-2} day^{-1}, and there were no difference between 20a and 100a because the GWP of the CO_2 will not change with the time.

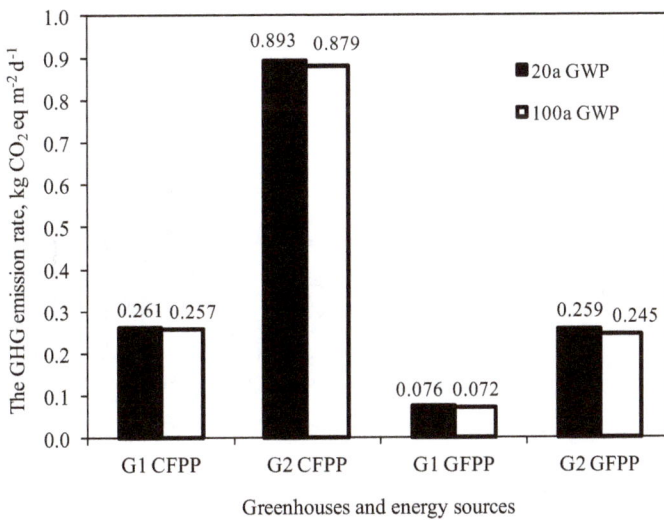

(a) The shallow geothermal energy

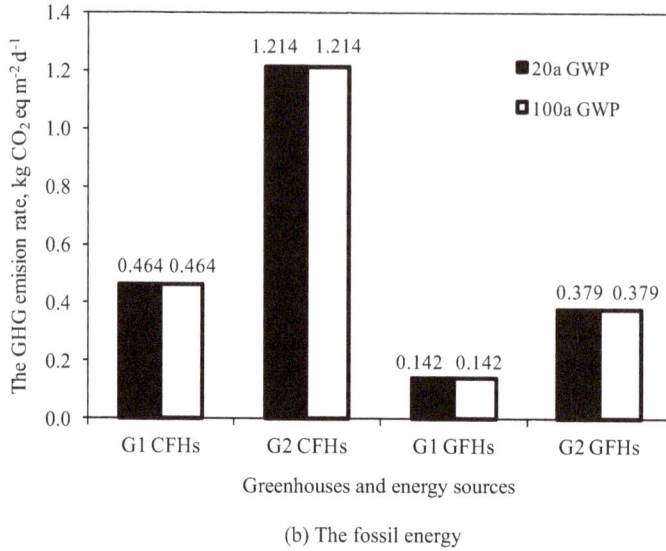

(b) The fossil energy

Figure 6. Standardized GHG emissions (GWP) for different greenhouse types and energy sources.

3.4. Emissions inventory in Beijing, China

According to the areas of the G1 and G2 in Beijing that require assisted heating (6000 ha Chinese solar greenhouses-G1 and 1000 ha multispan greenhouses-G2), the total GHG emissions from greenhouses heating with the CFHs or GFHs were quantified as 5238 or 2294 Mt CO_2 eq. in Beijing, and there is no difference between 20a and 100a GWP (Figure 7).

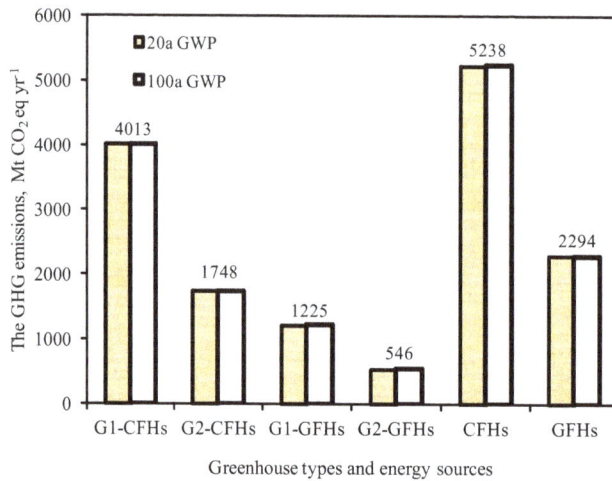

Figure 7. Total GHG emissions from heating greenhouses with fossil energies in Beijing.(G1-CFHs, heating all the G1-type greenhouses in Beijing with CFHs; G2-CFHs, heating all the G2-type greenhouses in Beijing with CFHs; G1-GFHs, heating all the G1-type greenhouses in Beijing with GFHs; G2-GFHs, heating all the G2-type greenhouses in Beijing with GFHs).

The total GHG emissions from heating greenhouses in Beijing with the GSHPs were quantified as 1658 and 2909 Mt CO_2 eq., based on 20a GWP or 1619 and 2839 Mt CO_2 eq. based on 100a GWP (Figure 8). The GHG emissions from heating G1-type greenhouses are higher than heating G2-type of greenhouses due to the large area of the G1 built and used in Beijing during 2007–2008 winter.

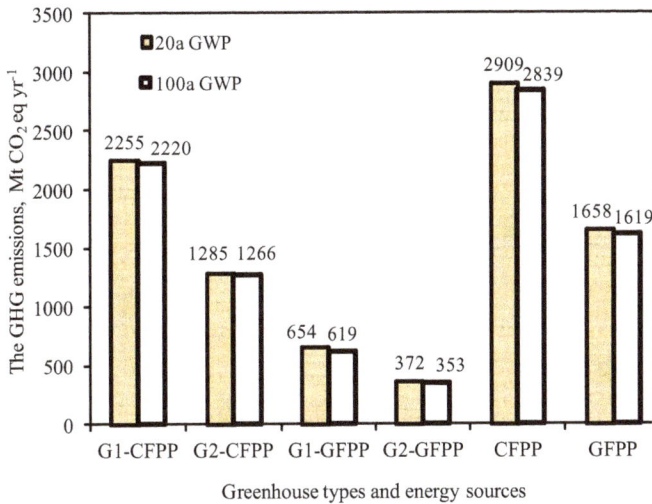

Figure 8. Total GHG emissions from heating greenhouses with the GSHPs in Beijing (G1-CFPP, heating all the G1-type greenhouses in Beijing with the GSHPs be powered with the electricity generated in CFHs; G2-CFPP, heating all the G2-type greenhouses in Beijing with the GSHPs be powered with the electricity generated in CFHs; G1-GFPP, heating all the G1-type greenhouses in Beijing with the GSHPs be powered with the electricity generated in GFHs; G2-GFPP, heating all the G2-type greenhouses in Beijing with the GSHPs be powered with the electricity generated in GFHs).

Applying the GSHPs to heat G1 and G2 with the electricity from the CFPP, the equivalent CO_2 emissions were 43% and 44% lower than directly burning coal with the CFHs but were 46.4% and 44.2% higher than the GFHs that burning natural gas. However, when using the GFPP generated electricity to run the GSHPs, the equivalent CO_2 emissions would be 83.5% and 83.8% lower than directly burning coal with the CFHs and were 45.9% and 48.1% lower than the GFHs that burning natural gas.

3.5. Uncertainty evaluation

It was assumed that all the solar greenhouse and multispan greenhouses with the same heating rate of G1 and G2 in this study, which would lead to errors due to the varying structures and materials in different greenhouses. For the solar greenhouses with improved wall materials and structures, the heat loss and heating rate would be lower [31]. Besides, heat-preserving technologies, such as multilayer aluminum foil heat reflecting materials, would have lower heat loss and heating rate than G2 in this study [3].

The shallow geothermal heat used in the GSHPs came from the groundwater (14°C), which has different GHG emissions from the borehole or U-tube-based HPs [19, 32]. Besides, the leaking factor of R22 was assumed to be 2% per year based on European studies, which may be changing with the change of the pump unit and maintenance of the system.

The GHG emissions calculated in this study are based on the real heating quantity required by G1 and G2 during winter heating, and the cycle of the SGE from extraction, enhance, and greenhouse heating was considered, the analysis can be considered as a partial life cycle assessment (LCA). However, a full LCA analysis could be applied to account the GHG emissions from greenhouse constructing with different materials, the transportation of the coal or natural gas for the location of the greenhouses, and the plants cultivated in the greenhouses [33-35].

In this study, we assumed that all the G1-type Chinese solar greenhouses would need additional heating in calculating the GHG emissions. However, novel structures and materials were applied for building Chinese solar greenhouses in Beijing in recent years [36], which improved the heat-preserving capacity of the greenhouse so that heating was not required in winter time. Therefore, the GHG emissions from heating Chinese solar greenhouse could be lower than the amount calculated in this study.

4. Summary

The unit electricity consuming rate of the GSHPs were 0.15 and 0.53 kWh m^{-2} d^{-1} for heating the Chinese solar greenhouse (G1) and multispan greenhouse (G2) or expressed as 1500 and 5300 kWh ha^{-1} d^{-1} in Beijing. The 20a and 100a GWPs of the GHG emissions from the GSHPs heating for G1 and G2 were 0.076–0.893 and 0.072–0.879 g CO_2 eq. m^{-2} d^{-1}, respectively.

The total GHG emissions from heating greenhouses in Beijing with the GSHPs were quantified as 1658–2909 Mt CO_2 eq. Among different stages of the SGE flow, most GHG emissions (66%) happened at the stage of SGE promotion due to the higher consumption of electricity in compressors.

The total GHG emissions from greenhouses heating with the CFHs or GFHs were quantified as 5238 and 2294 Mt CO_2 eq. in Beijing, respectively. Applying the GSHPs to heat G1 and G2 with the electricity from the CFPP, the equivalent CO_2 emissions were 43% and 44% lower than directly burning coal with the CFHs but were 46.4% and 44.2% higher than the GFHs that burning natural gas. However, when using the GFPP-generated electricity to run the GSHPs, the equivalent CO_2 emissions would be 83.5% and 83.8% lower than directly burning coal with the CFHs and were 45.9% and 48.1% lower than the GFHs that burning natural gas.

The glass-covered G2 consumed more heating energy than G1 during the heating period. This demonstrated that the Chinese solar greenhouse design had better heat preservation than the glass greenhouse. Besides, novel structures and materials applied for building Chinese solar greenhouses in Beijing could further reduce the GHG emissions from heating.

Acknowledgements

The study was sponsored by the "Beijing Natural Science Foundation (6132011)," "Young Researcher Foundation (QNJJ201212)" in Beijing Academy of Agricultural and Forestry Sciences, "Earmarked Fund for Modern Agro-industry Technology Research System (CARS-25-D-04)," and "Twelve-Five-Year National Science and Technology Support Program (2011BAD12B01)."

We also thank the support from the program of "Promotion network of research and application services on vegetables varieties (KJCX20140416)".

Author details

Lilong Chai[1,2*], Chengwei Ma[3], Baoju Wang[1,4], Mingchi Liu[3,4] and Zhanhui Wu[3,4]

*Address all correspondence to: lchaipurdue@gmail.com

1 National Engineering Research Centre for Vegetables, Beijing Academy of Agriculture and Forestry Sciences, Beijing, P.R. China

2 Department of Agricultural and Biosystems Engineering, Iowa State University, Ames, USA

3 College of Water Resource and Civil Engineering, China Agricultural University, P.R. China, Beijing, P.R. China

4 Key Laboratory of Urban Agriculture (North), Ministry of Agriculture, P.R. China, Beijing, P.R. China

References

[1] Bot G.P.A. (2001). Developments in indoor sustainable plant production with emphasis on energy saving. Computers and Electronics in Agriculture, 30(1-3), 151-165.

[2] Torrellas M., Assumpcio A., Juan I.M. (2013). An environmental impact calculator for greenhouse production systems. Journal of Environmental Management, 118, 186-195.

[3] Ma C., Miao X. (2005). Agricultural Bio-environment Engineering. China Agriculture Press, Beijing, China.

[4] Dickson M.H., Fanelli M. (2013). Geothermal Energy: Utilization and Technology. Routledge.

[5]   Intergovernmental Panel on Climate Change (IPCC). (2006). Guidelines for National Greenhouse Gas Inventories. Institute for Global Environmental Strategies (IGES) for the IPCC, Kanagawa, Japan.

[6]   Bayer P., Saner D., Bolay S., Rybach L., Blum P. (2012). Greenhouse gas emission savings of ground source heat pump systems in Europe: a review. Renewable and Sustainable Energy Reviews, 16(2), 1256-1267.

[7]   Self S.J., Reddy B.V., Rosen M.A. (2013). Geothermal heat pump systems: status review and comparison with other heating options. Applied Energy, 101, 341-348.

[8]   Omer A.M. (2008). Ground-source heat pumps systems and applications. Renewable and Sustainable Energy Reviews, 12(2), 344-371.

[9]   Sonneveld P.J., Swinkels G.L.A.M., Bot G.P.A., Flamand G. (2010). Feasibility study for combining cooling and high grade energy production in a solar greenhouse. Biosystems Engineering, 105(1), 51-58.

[10] Blum P., Campillo G., Kölbel T. (2011). Techno-economic and spatial analysis of vertical ground source heat pump systems in Germany. Energy, 36(5), 3002-3011.

[11] Hähnlein S., Bayer P., Ferguson G., Blum P. (2013). Sustainability and policy for the thermal use of shallow geothermal energy. Energy Policy, 59, 914-925.

[12] Adaro J.A., Galimberti P.D., Lema A.I., Fasulo A., Barral J.R. (1999). Geothermal contribution to greenhouse heating. Applied Energy, 64 (1-4), 241-249.

[13] Ozgener O., Hepbasli A. (2005). Experimental investigation of the performance of a solar-assisted ground-source heat pump system for greenhouse heating. International Journal of Energy Research, 29(3), 217-231.

[14] Underwood C.P., Spitler J.D. (2007). Analysis of vertical ground loop heat exchangers applied to buildings in the UK. Building Services Engineering Research & Technology, 28(2), 133-159.

[15] Ozgener O., Hepbasli A. (2007). A review on the energy and exergy analysis of solar assisted heat pump systems. Renewable & Sustainable Energy Reviews, 11(3), 482-496.

[16] Sethi V.P., Sharma S.K. (2008). Survey and evaluation of heating technologies for worldwide agricultural greenhouse applications. Solar Energy, 82(9), 832-859.

[17] Tong Y., Kozai T., Nishioka N., Ohyama K. (2010). Greenhouse heating using heat pumps with a high coefficient of performance (COP). Biosystems Engineering, 106(4), 405-411.

[18] Chai L., Ma C., Ni J.-Q. (2012). Performance evaluation of ground source heat pump system for greenhouse heating in northern China. Biosystems Engineering, 2012, 111(1), 107-117.

[19] Heikkila K. (2008). Environmental evaluation of an air-conditioning system supplied by cooling energy from a bore-hole based heat pump system. Building and Environment, 43(1): 51-61.

[20] Han R., Li W., Ai W., Song Y., Ye D., Hou W. (2010). The climatic variability and influence of first frost dates in northern China. Acta Geographica Sinica, 65 (5), 525-532.

[21] Luo W., de Zwart H. F., Dai J., Wang X., Stanghellini C., Bu C. (2005). Simulation of greenhouse management in the subtropics. Part I: Model validation and scenario study for the winter season. Biosystems Engineering, 90(3), 307-318.

[22] Tong G., Christopher, D.M., Li, B. (2009). Numerical modelling of temperature variations in a Chinese solar greenhouse. Computers and Electronics in Agriculture, 68(1), 129-139.

[23] Tong G., Wang T., Bai Y., Liu W. (2003). Heat transfer property of the wall in solar greenhouse. Transactions of the CSAE, 19(3), 186-189 (in Chinese with English abstract).

[24] Bartzanas T., Tchamitchian M., Kittas C. (2005). Influence of the heating method on greenhouse microclimate and energy consumption. Biosystems Engineering, 91(4), 487-499.

[25] Beijing Statistical Bureau (2008). Beijing Statistical Yearbook 2008. Chinese Statistical Press, Beijing.

[26] Li Z.M., Shen J., Wang Z., Gao L H., Chen Q.Y., Guo Y.X. (2011). Production efficiency analysis of solar greenhouse and plastic big-arch shelter in Beijing. China Veget, 22, 13-19.

[27] Chai L., Ma C., Zhang Y., Wang M., Ma Y., Ji X. (2010). Energy consumption and economic analysis of ground source heat pump used in greenhouse in Beijing. Transactions of the CSAE, 26(3), 249-254 (in Chinese with English abstract).

[28] Zhang J. (1999). Simplified Handbook for the Air Conditioning Professional. Beijing Industry Press, Beijing, China.

[29] Liu Y., Ma Z., Zou P. (2002). Heating Ventilating and Air Conditioning. China Building Industry Press, Beijing, China.

[30] Jones J C (2008). Atmosphere Pollution. www.bookboon.com

[31] Chai L., Ma C., Ji X., Yang R., Zhou Z., Bu Y. (2007). Performances analysis on the energy-saving materials utilized in building solar greenhouses in Beijing. Journal of Agricultural Mechanization Research, 8, 17-21 (in Chinese with English abstract).

[32] Blum P., Campillo G., Münch W., Kölbel T. (2010). CO_2 savings of ground source heat pump systems–a regional analysis. Renewable Energy, 35(1), 122-127.

[33] Russo G., Scarascia Mugnozza G. (2004). LCA methodology applied to various typology of greenhouses. In International Conference on Sustainable Greenhouse Systems-Greensys 2004, September, 691: 837-844.

[34] Torrellas M., Antón A., Ruijs M., Victoria N. G., Stanghellini C., Montero J.I. (2012). Environmental and economic assessment of protected crops in four European scenarios. Journal of Cleaner Production, 28, 45-55.

[35] Xing S., Xu Z., Jun G. (2008). Inventory analysis of LCA on steel-and concrete-construction office buildings. Energy and Buildings, 40(7), 1188-1193.

[36] Chai L., Wang B., Liu M., Wu Z., Xu Y. (2014). Dual-roof solar greenhouse—a novel design for improving the heat preserving capacity in Northern China. Natural Resources, 5(12), 681.

6

Greenhouse Gas Emissions – Carbon Capture, Storage and Utilisation

Bernardo Llamas, Benito Navarrete, Fernando Vega, Elías Rodriguez, Luis F. Mazadiego, Ángel Cámara and Pedro Otero

Additional information is available at the end of the chapter

Abstract

According to the recent information, CO_2 concentration in the atmosphere reached 402 ppm at the beginning of 2016. On the other hand, fossil fuels remain as the major source to produce energy. The International Energy Agency estimate that those fuels will remain as the most used source during coming decades.

Carbon capture and storage technology is the most promising technology to significantly decrease CO_2 emissions. Nevertheless, it may be possible to use CO_2 as a raw material for other industrial uses. In this chapter, authors explain both ways to decrease CO_2 emissions.

Keywords: CCS technology, CO_2 capture technologies, CO_2 storage, CO_2 uses, macrofouling

1. Introduction

The Fifth Assessment Report from the Intergovernmental Panel on Climate Change states that human influence on the climate system is clear [1]. The CO_2 concentration in the atmosphere is continuously growing. The latest value is 402.52 ppm (January 2016, Mauna Loa Observatory), which is 2 pmm higher than the value registered in January 2015 [1].

Carbon capture and storage (CCS) is a way of 'decarbonising' fossil fuel power generation. It involves capturing carbon dioxide (CO_2) emitted from high-producing sources, transporting it and storing it in secure geological formations deep underground, to mitigate the effect of greenhouse emissions on climate change [2].

The transported CO_2 can also be reused in processes such as enhanced oil recovery (EOR) or in the chemical industry, a process sometimes known as carbon capture and utilisation (CCU). CCS can be applied to fossil fuel power plants (coal and gas-fired power stations) and to industrial CO_2-emitting sources such as oil refineries or cement, chemical and steel plants. Rather than being a single technology, CCS is a suite of technologies and processes. While some of these have been operated successfully for decades, progress in applying large-scale CCS to power generation globally has been slow (Figure 1).

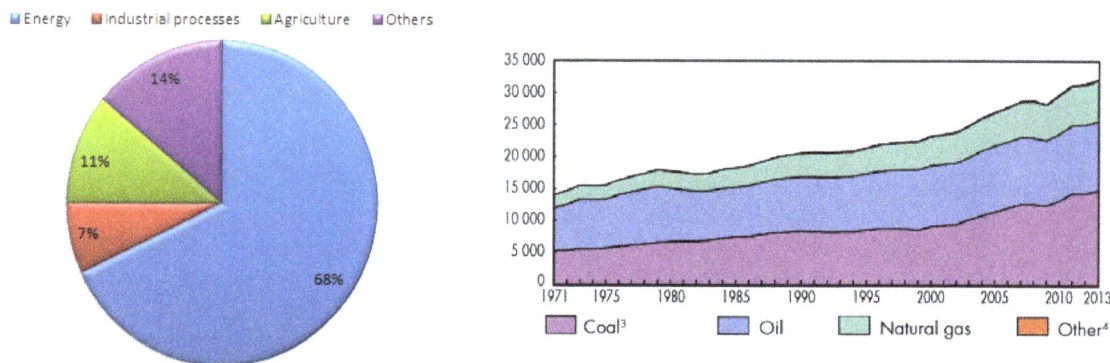

Figure 1. Shares of global anthropogenic greenhouse gas emissions (GHG) and world CO_2 emissions from fuel combustion by fuel (Mt of CO_2) [3, 4].

Carbon capture and storage (CCS) is likely to be a crucial part of the least-cost path to decarbonisation. It can provide a back-up role for variable renewables and help to manage swings in demand. CCS also has a crucial role in decarbonising heavy industry where there are limited options, and in the longer term would help to maximise the emission reduction obtained from scarce supplies of sustainable bioenergy as well as opening up other decarbonisation pathways.

The European Commission has also emphasised that 'CCS may be the only option available to reduce direct emission from industrial processes at the large scale needed in the longer term'.

In this chapter, authors review the carbon capture, storage technology (including the CO_2 transport through pipeline), and CO_2 utilisation technologies.

2. CO_2 capture

This process consists of the separation of CO_2 from flue gas produced during the combustion of fossil fuels and can be applied to large flue gas stationary sources as thermal power stations and industrial processes.

Current CO_2 capture technology (first generation) is adapted from gas separation processes already in industrial use. There are several technologies and strategies to capture CO_2 from stationary sources: pre-combustion, post-combustion and oxy-fuel (Figure 2).

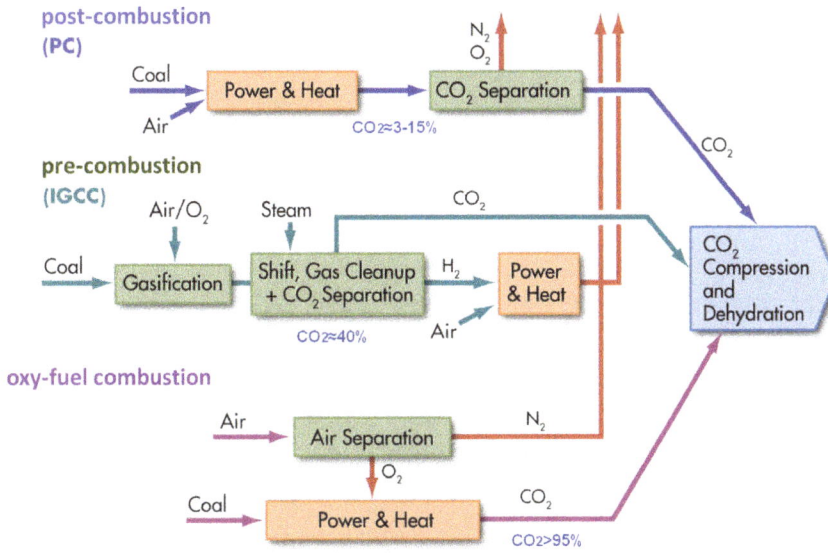

Figure 2. Summary of CO_2 capture technologies (adapted from IPCC) [2].

3. First generation of capture technologies

3.1. Post-combustion capture

Post-combustion capture follows the conventional application of a specific purification unit applied for a particular pollutant removal (CO_2 in this case). Figure 3 illustrates a typical block diagram of the post-combustion process that offers a great feasibility and versatility in terms of operating conditions and process integration.

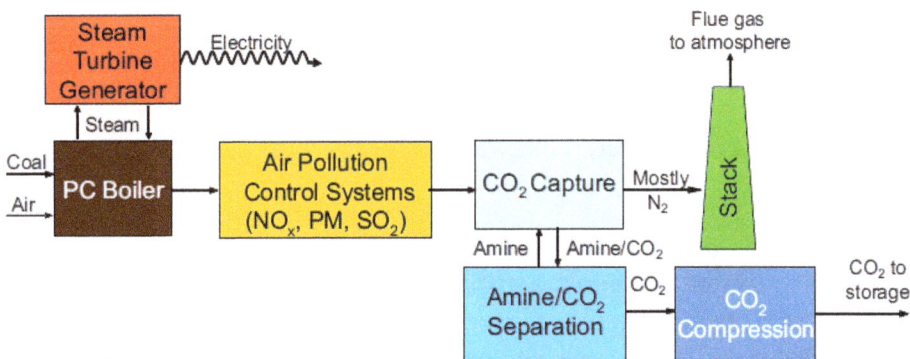

Figure 3. Simplified scheme of a fossil-fuel power plant using a post-combustion capture unit [5].

CO_2 concentration in the flue gas from a combustion process varies from 4 to14% in natural gas and coal-power plants, while other industries such as cement, iron and steel and petrochemical produce flue gas ranging between 14 and 33%. The key drawbacks hindering the large-scale implementation of this technology lies in the large volume of gas that should be treated and the low CO_2 concentration of the flue together with high energy requirements, mainly related to CO_2 desorption process. The presence of large amounts of dust, O_2, SOx, NOx and trace pollutants such as Hg and the relatively high temperature of the flue gas, typically between 120 and 180°C, are also design challenges that have significant impact on the capture costs.

The technologies currently available for post-combustion capture are classified into five main groups: absorption, adsorption, cryogenics, membranes and biological separation. The most mature and closest to market technology and so, the representative of first generation of post-combustion options, is capture absorption from amines.

3.2. Chemical absorption from amines

Post-combustion capture using chemical absorption by aqueous alkaline amine solutions has been used for CO_2 and H_2S removal from gas-treating plants for decades [6]. Amines react rapidly, selectively and reversibly with CO_2 and can be applied at low CO_2 partial pressure conditions. Amines are volatile, cheap and safe in handling. They show several disadvantages as they are also corrosive and require the use of resistant materials. Furthermore, amines form stable salts in the presence of O_2, SO_X and other impurities such as particles, HCl, HF and organic and inorganic Hg trace compounds that extremely constrain the content of those compounds in the treated gas.

The most widely used amine is monoethanolamine (MEA), which is considered as a benchmark solvent because of its high cyclic capacity, significant absorption-stripping kinetic rates at low CO_2 concentration and high solubility in water. Some other amine-based solvents such as diethanolamine (DEA), triethanolamine (TEA), diglycolamine (DGA), N-methyldiethanolamine (MDEA), piperazine (PZ), 2-amino-2-methyl-1-propanol (AMP) and N-(2-aminoethyl)piperazine (AEP) have also traditionally been utilised.

A typical chemical absorption scheme is shown in Figure 4. A low CO_2 concentrated flue gas is introduced in the absorber in crosscurrent with lean solvent from the stripper at 50–55°C and ambient pressure. CO_2 reacts with amines in the absorber according to the overall reaction:

$$CO_2 + 2R_1R_2NH \leftrightarrow R_1R_2NCOO^- + R_1R_2NH_2^+ \tag{1}$$

As CO_2 is absorbed, rich amine from the absorber bottom is fed into a cross-exchanger with lean amine before it is introduced into the stripper. The stripping temperature varies between 120 and 150°C, and the operating pressure reaches up to 5 bar. A water saturated CO_2 stream is released from the top and is subsequently ready for transport and storage, while lean amine leaving the stripper is pumped back into the absorber.

The high energy penalty related to amines regeneration (a high-intensive energy process because of the stripper operating conditions and solvent used) and solvent degradation are the issues most hindering a large deployment of this technology.

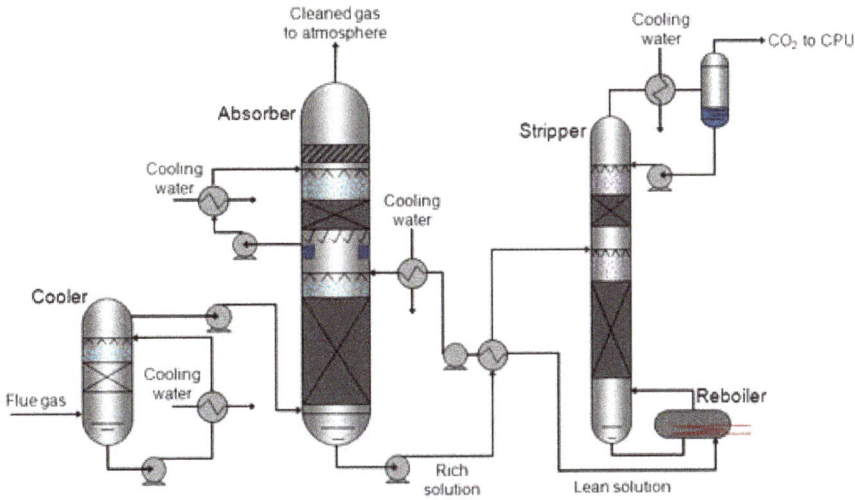

Figure 4. Diagram of a conventional CO_2 capture process using amine-based chemical absorption.

3.3. Pre-combustion capture

In pre-combustion CO_2 capture, CO_2 separation occurs prior to fuel combustion and power generation (Figure 5). The fuel reacts at high temperature and pressure with either oxygen or/ and steam under sub-stoichiometric conditions, and thereby a gas stream primarily composed of CO and H_2 is obtained. This CO/H_2 gas mixture is commonly known as *synthesis gas* or **syngas**.

In general, steam is utilised in case fuel is solid, namely gasification, whereas sub-stoichiometric oxygen is used with liquid and gaseous fuels. Both reactions occur at elevated temperature (1,400°C) and pressure (3–7 Mpa), as seen in Equations 2 and 3.

Steam reforming:

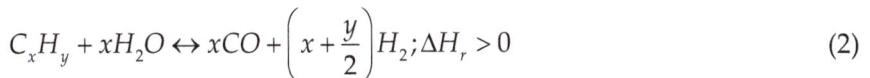

$$C_xH_y + xH_2O \leftrightarrow xCO + \left(x + \frac{y}{2}\right)H_2; \Delta H_r > 0 \tag{2}$$

Partial oxidation:

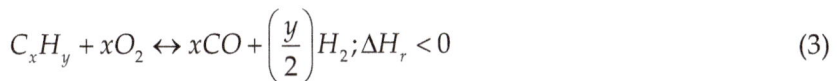

$$C_xH_y + xO_2 \leftrightarrow xCO + \left(\frac{y}{2}\right)H_2; \Delta H_r < 0 \tag{3}$$

Steam reforming needs a secondary fuel to provide the energy supply necessary for the reaction that occurs and a catalysts to improve the kinetic of this process. In Equation (3), the primary fuel is partially oxidised by a limited amount of oxygen. Partial oxidation produces less H_2 per fuel unit than stream reforming, but the kinetic reaction is faster, it requires smaller reactors and neither catalyst nor energy supply from a secondary fuel.

Once particulate matter is removed, the syngas passes through a two stages catalytic reactor, where CO reacts with steam to produce CO_2 and further yield H_2: *water-gas-shift (WGS) reaction*.

WGS reaction:

$$CO + H_2O \leftrightarrow CO_2 + H_2; \Delta H_r = -41\,kJ/mol \qquad (4)$$

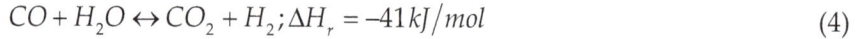

The syngas resulted is mainly composed of CO_2, ranging from 15 to 40%v/v, and H_2 at elevated pressure from which CO_2 can be easily separated by a physical absorption mechanism and then CO_2 can be easily released by simply dropping pressure.

Before the syngas from WGS reactor is separated into its primary components, the sulphur compounds, mainly in COS and H_2S form, are removed to avoid its emission to the atmosphere. Sulphur is then recovered in either as solid in a Claus plant or as sulphuric acid.

The sulphur-free syngas has a high CO_2 concentration and an elevated pressure (2–7 MPa), thus making physical absorption highly recommended for CO_2 separation, although adsorption process such as pressure swing adsorption (PSA) is also utilised.

The remaining nearly pure H_2 stream could be burned in a combined cycle power plant to generate electricity, but H_2 turbines require further development. Power fuel cells and transportation fuels are alternative options for using H_2 in the future, currently under development.

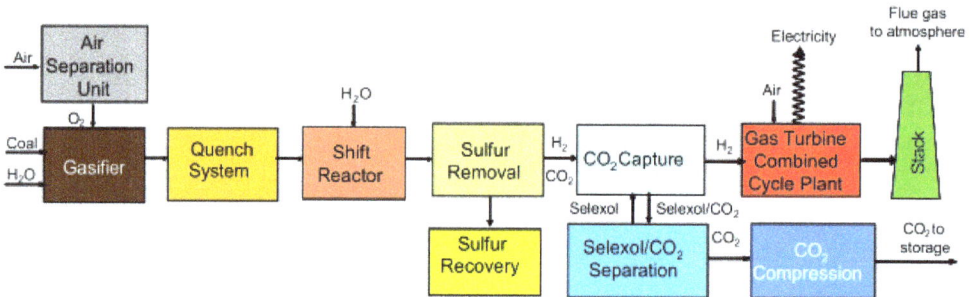

Figure 5. Simplified scheme of an integrated gasification combined cycle (IGCC) coupled with a pre-combustion CO_2 capture and storage unit using a physical absorption process [5].

3.4. Oxy-combustion capture

Oxy-combustion or oxy-fuel capture is considered as one of the most promising CCS technologies that would be economically competitive in fossil-fuel power plants and industrial facilities. It has been developed for both new designs and retrofitting of existing plants, although it is best adapted to newly designed power plants. A basic process flow diagram is given in Figure 6. Oxy-combustion technology is based on the use of high purity O_2 as oxidiser in an O_2/CO_2 mixture instead of air during the combustion process. It has been first proposed for coal boilers and gas turbines but can be applied to any type of fossil fuel utilised for thermal power production. As burning with O_2 at high concentration can produce high flame temperatures in the boiler, part of the exhaust gas from the boiler, mainly CO_2 and water vapour (FGR flue gas recirculation stream), is recycled to control temperatures to levels compatible with available boiler materials. The flue gas obtained from this system consists mainly of CO_2 and H_2O and are accompanied by minor quantities of N_2, SOx, NOx, Ar and Hg. Water can be easily removed by condensation, producing a highly CO_2 concentrated flue gas. The CO_2 content varies from 70 to 95%v/v, depending on the process configuration, air in-leakages, fuel characteristics and the purity of O_2.

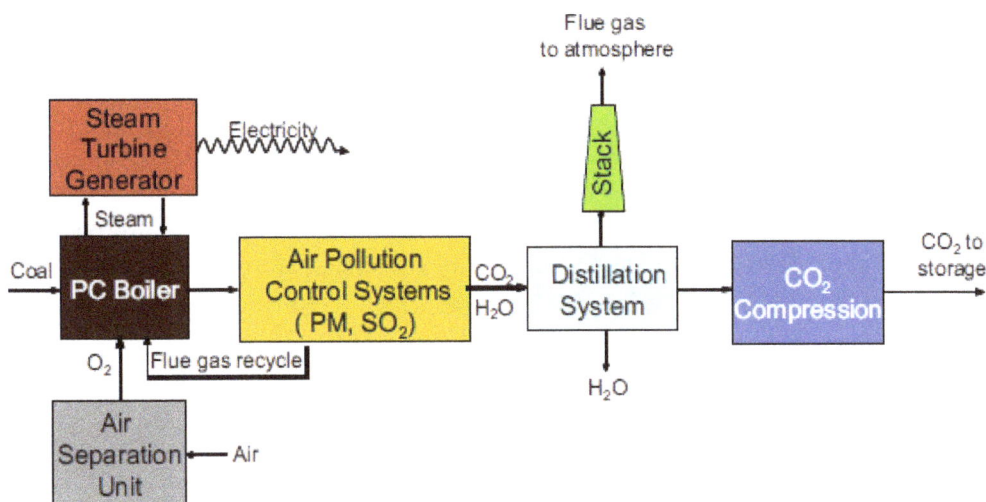

Figure 6. A simplified scheme of a fossil-fuel power plant based on the oxy-combustion concept [5].

Oxy-combustion requires large amounts of high purity (95–99%) O_2 for power production. A typical 500 MWe fossil-fuel power plant would need 9,000–10,000 t/d to operate under oxy-combustion conditions [7]. Currently, cryogenic distillation is the only available technology that can supply those amounts of O_2. An air separation unit (ASU) can provide around 4,500–7,000 t/d of oxygen, while other alternative technologies such as vacuum pressure swing adsorption (VPSA) units and membranes can only produce one order of magnitude below ASU production.

The ASU would consume up to 60% of the total electricity required for carbon capture and reduces the overall efficiency of the power plant by about 7–9%, reaching up to 15% in some cases. Furthermore, the availability and rapid response of the ASU to load changes have been noted as crucial challenges for the global oxy-combustion plant operation and feasibility. New technologies for O_2 production as ion transport membranes (ITM) or VPSA have shown promising results related to energy consumption, but the large amounts of O_2 required in power plant operation avoid currently its commercial deployment.

The CO_2 stream obtained from oxy-fuel combustion shows high levels of water vapour, sulphur compounds, N_2, O_2 and impurities such as mercury in the flue gas. NOx emission is low when compared with air combustion.

The CO_2 gas quality has significant impact on the capture cost by this technology, and uncertainties on the future regulatory requirements of CO_2 quality for its transport and storage has influence on the process configuration of the oxy-combustion plant, gas cleaning unit performance, overall CO_2 recovery capacity and on the energy requirements for CO_2 compression and purification.

4. Emerging technologies for CO_2 capture

The most promising emerging technologies applied to carbon capture are discussed in this section to complete the overview of the CO_2 capture technologies currently under research.

4.1. Chemical looping combustion

Chemical looping combustion (CLC) is a promising technology for fuel combustion, which can be beneficial in carbon capture applications. It is based on the use of an oxygen carrier, typically a metal oxide, to supply the O_2 needed for the fuel combustion process, producing a highly CO_2 concentrated exhaust gas. Iron, nickel, cobalt, copper, manganese and cadmium are commonly used as oxygen carriers in CLC.

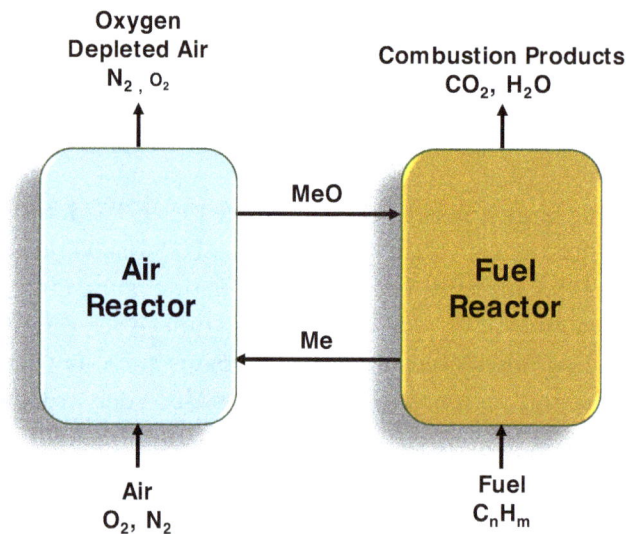

Figure 7. A simplified scheme of a chemical looping for oxy-combustion.

CLC consists of two fluidised bed reactors, namely reducer and oxidiser. In the reducer reactor, fuel is fed along with the metal oxide containing oxygen, which is transferred from the metal oxide to the reactor as the combustion occurs (Figure 7). A flue gas containing over 99%v/v of CO_2 can be obtained by a simply condensation stage because of the fact that the exhaust gas at the reducer outlet is primarily formed by CO_2 and water vapour. This stream is then sent to further compression and permanent storage.

Reducer:

$$\left(2n+m\right)M_yO_x+C_nH_{2m}\leftrightarrow\left(2n+m\right)M_yO_{x-1}+mO+nCO_2.\tag{5}$$

Oxidiser:

$$M_yO_{x-1}+\frac{1}{2}O_2\leftrightarrow M_yO_x+N_2+O_2\left(excess\right).\tag{6}$$

4.2. Hydrate-based separation

This separation approach is based on the hydrate formation from high pressure water in contact with the flue gas containing CO_2. Hydrates are crystalline under suitable low temperature and high pressure conditions. A pure CO_2 stream is then obtained as CO_2 is released from the hydrates, achieving up to 99% of CO_2 recovery.

4.3. Calcium looping

Calcium looping is based on the reversible reaction between CaO and CO_2 to form calcium carbonate.

Calcium looping consists of two fluidised bed reactors, namely carbonator and calciner. In the carbonator, primary fuel is burned and CaO reacts with the CO_2 formed from the fuel combustion following the reaction seen in Equation (7). Carbonator temperature is within 650–700°C, depending on the system pressure.

Carbonator:

$$CaO(s)+CO_2(g)\leftrightarrow CaCO_3(s).\tag{7}$$

Calciner:

$$CaCO_3(s)\leftrightarrow CaO(s)+CO_2(g).\tag{8}$$

$CaCO_3$ is then heated by secondary fuel combustion in the calciner. CaO is regenerated and CO_2 is released for storage according to the reaction in Equation (8). The calciner temperature can reach 900°C, depending on the CO_2 partial pressure.

This technology shows benefits for carbon capture. Limestone is cheap and widely available, and there is a potential for process integration, which can lead to low energy penalties, i.e., heat released from carbonisation can be utilised in a steam cycle or the heat used in the calciner reactor can be recovered in the carbonation process.

4.4. Partial oxy-combustion

The energy consumption required for solvent regeneration and high purity oxygen production is the major drawback of post-combustion and oxy-combustion technologies. A new hybrid concept has been proposed to reduce the energy requirements associated with CO_2 capture step combining a partial oxy-fuel combustion (using oxygen-enriched air instead of high purity oxygen as oxidiser) and a CO_2 separation process treating a flue gas with a higher CO_2 concentration than in conventional air combustion (Figure 8).

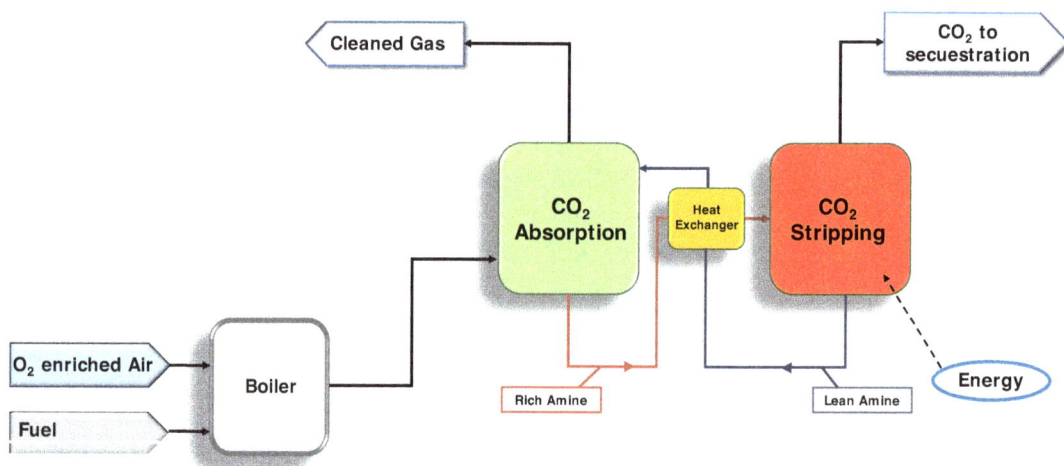

Figure 8. A simplified scheme of a partial oxy-combustion plant.

The combination of a less-constrained ASU for oxygen production and a carbon capture process using membranes instead of amine solvents can conduce to a minimal energy requirement associated with an oxygen purity ranging between 0.5 and 0.6 molar fraction.

4.5. Biological CO_2 capture

Biological CO_2 capture from a gas mixture is based on natural reactions of CO_2 with living organism, mainly enzymes, generally proteins and (micro)algae. Enzymes catalyse CO_2 chemical reaction and enhance CO_2 absorption rate in water. Enzymes can be also immobilised at the gas-liquid interface to promote CO_2 dissolution from the bulk gas. In this sense, carbonic anhydrase enzyme supported in a hollow fibre with liquid membrane has been reported as a potential method applied to CO_2 capture, achieving up to 90% CO_2 capture associated with low energy requirements in the regeneration process at laboratory-scale experiments. Carbonic anhydrase promotes carbonic acid formation from dissolved CO_2 and enhances CO_2 absorption from gas phase using and extremely low CO_2/enzyme ratio. CO_2 separation using enzymes must incorporate a tailored regeneration process to produce a high concentrated CO_2 exhaust stream. Membrane boundary, fouling, long-term operation and pore wetting are identified as the most relevant technical issues to be addressed before the scale-up of this CO_2 capture approach.

The use of algae is also considered a promising CO_2 capture option among natural occurring reactions. Algae consume CO_2 through photosynthesis mechanism. The use of algae in CO_2 capture would avoid subsequent CO_2 compression and storage stages, but there are some key issues that must be addressed for its large-scale deployment. In fact, algae require excessive amount of water and large gas-liquid interface surfaces that drastically limit their application in carbon capture. Algae are also highly susceptive to changes in operating conditions and to the presence of impurities such as vanadium and nickel.

4.6. Ionic liquid absorption

Significant progress has been made in the application of ionic liquids (ILs) as alternative solvents to CO_2 capture because of their unique properties such as very low vapour pressure, a broad range of liquid temperatures, excellent thermal and chemical stabilities and selective dissolution of certain organic and inorganic materials. ILs are liquid organic salts at ambient conditions with a cationic part and an anionic part.

ILs have the potential to overcome many of the problems of associated with current CO_2 capture techniques. ILs are particularly applicable in absorption of CO_2 while effectively avoiding the loss of sequestering agents. Other advantage of ILs is that they can be combined into polymeric forms, increasing the CO_2 sorption capacity compared with other ILs and conventional solvents and greatly facilitates the separation and ease of operation.

5. CO_2 transport

Currently there are more than 6,500 km of CO_2 pipelines worldwide. Most of them deliver CO_2 to EOR operations in the United States, but there is also a growing number under development for CO_2 storage projects

The relative development of the infrastructure to transport CO_2 is still in its early stages. This is reflected by the low number of existing infrastructures developed to transport CO_2 from stationary sources into geological structures. Table 1 provides an overview of the current developments for CO_2 transportation globally. All of these examples have been developed in relation to the EOR technique, where the CO_2 source is found mainly in natural reserves. In Europe, only a few projects are in operation, but there are plans to deploy an extended CO_2 pipeline network along Europe to optimise CO_2 storage structures.

These examples may be used to study CO_2 conditions; in addition, many CO_2 pipeline projects are based on well-known designs and materials commonly used in natural gas pipeline specifications. The most profitable way to transport CO_2 is in its dense phase [9].

To avoid two phases, it has been suggested that the most efficient way to transport CO_2 is as its supercritical phase [8, 9], which occurs at a pressure higher than 7.38 MPa and a temperature of more than 31.1 °C. To maintain these conditions, this type of transportation may require the use of booster stations in the pipeline layout to maintain the required pressure and temperature.

Pipeline	Location	Length (Km)	Diameter (inches)	Estimated maximum (10^6t/year)
Cortez	US	808	30	23.6
Sheep Mountain	US	656	NA	11.0
Bravo	US	351	20	7.0
Dakota Gasification/Weyburn	US/Canada	328	14	2.6
Choctaw	US	294	20	7.0
Bairoil	US	258	NA	23.0
Central Basin	US	230	16	4.3
Canyon Reef Carriers	US	224	16	4.3
Comanche Creek	US	193	6	1.3
Centerline	US	182	16	4.3
Delta	US	174	24	11.4
Snohvit	Norway	153	NA	0.7
Borger	US	138	4	1.0
Coffeyville	US	112	8	1.6
OCAP	The Netherlands	97	NA	0.4
Beaver Creek	US	85	NA	NA
Anton Irish	US	64	8	1.6
El Mar	US	56	6	1.3
Chaparral	US	37	6	1.3
Doliarhide	US	37	8	1.6
Lacq	France	27	NA	0.1
Adair	US	24	4	1.0
Cordona Lake	US	11	6	1.3

Table 1. Current CO_2 pipelines. The first long distance CO_2 pipeline was in the 1970s. Main utilisation of the natural & anthropogenic CO_2 is EOR activities [8].

Material selection should be compatible with all states of the CO_2 stream. They should be defined to prevent corrosion and maximum material stress. In addition, eligible materials need to be qualified for the potential low temperature conditions that may occur during a pipeline depressurisation situation.

The design of a pipeline should meet the requirements set by appropriate regulations and standards. CO_2 pipelines shall be designed according to applicable regulatory requirements. The Recommended Practice for Design and Operation of CO_2 refers to the following pipeline standards: ISO 13623:2009, DNV-OS F101:2012 and ASME B31.4 or ASME B31.8.

Usually CO_2 pipelines are designed using existing national standards for gas and liquid transportation pipes, while additional CO_2 specific design issues are taken into consideration by the pipeline construction/operation companies to guarantee the reliable and safe operation of a given pipeline.

The use of carbon steels (e.g., with API X-60 and X-65) for the transportation of CO_2 streams has been ongoing for more than 30 years, as required in EOR projects. During the 2002–2008 period, 18 incidents were reported with no fatalities and/or injuries.

	Range
Length (km)	1.09–808
External diameter (mm)	152–921
Wall thickness (mm)	5.2–27
Capacity design (Mt/y)	0.06–28
Pressure min (bar)	3–151
Pressure max (bar)	21–200
Compressor capacity (MW)	0.2–68

Table 2. Summary of the current parameters considered in the CO_2 transport phase.

The cost of pipeline transportation will be determined by the pipeline route, in which physical and social geography will be crucial conditions.

The three major cost elements for pipelines are (1) construction costs (e.g., materials, labour, booster station, if needed, and others), (2) operation and maintenance costs (e.g., monitorisation, maintenance, energy costs) and (3) other costs (design, insurance, fees, and right-of-way).

6. CO_2 storage

At present, there are three possible geological structures that may be considered for CO_2 storage: depleted hydrocarbon and production, deep saline aquifers, and coal seams.

6.1. Depleted hydrocarbon fields

The CO_2 can be stored in supercritical conditions, rising by buoyancy and can be physically held in a structural or stratigraphic trap, the same way as the natural accumulation of hydrocarbons occurs. The advantage of the capacity of containment system has been demonstrated by the retention of oil for millions of years. If the site is in production, it is used to increase the recovery of oil or gas (EOR recovery – enhanced oil, gas-enhanced recovery – EGR). These operations, EOR/EGR, provide an economic benefit that can offset the costs of the capture, transport and storage of CO_2.

6.2. Deep saline aquifers

They are the best options for storing large volumes of CO_2 because of its size and found more than 800 meters below the surface. The supercritical CO_2 is 30–40% less dense than typical saline water from these formations, which means that the CO_2 naturally rise by buoyancy through the reservoir until it is caught or becomes longer solution term. They require an impermeable cap rock to ceiling (shales or layers of evaporites) and a porous and permeable rock store (sandstone or limestone) that promotes the injection, migration and trapping.

6.3. Coal seams

CO2 in gaseous form is injected into the coalbed, 300 to 600 metre depth, and adsorbed on the matrix pores, releasing the existing CH_4 in the same (two molecules of CO_2 adsorbed by each CH_4 molecule that travels). This has led to the possibility of storing CO_2 in coal seams, while CH_4 recovered is valued. This technique is called 'enhanced coalbed methane production' (ECBM).

Coal properties (range, degree and permeability) determine the suitability of the site, either for storage or storage with only CH_4 recovery.

6.4. Site selection and exploration

Figure 9 represents a proposed work flow for any CO_2 storage project. It is possible to determine three mayor phases: pre-injection, injection and post-injection phases.

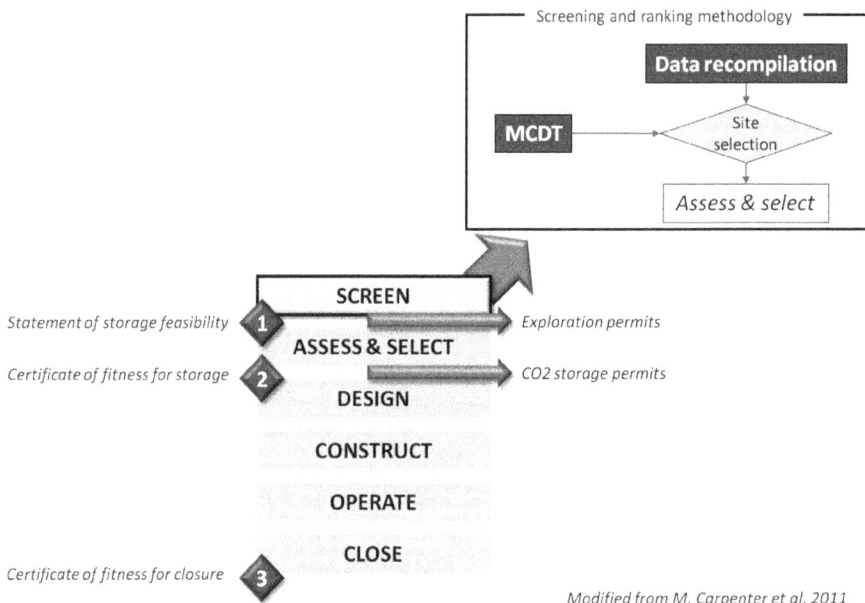

Figure 9. Work flow proposed for basin screening (Definition phase) [10].

In general, most of the areas that could be suitable for storing CO_2 are not well explored geologically. For this reason, pre-injection phase is crucial to decrease the inherent risk in subsurface exploration.

Screen phase could be differentiated by the data recompilation task and the multicriteria decision tool. It is integrated as a preliminary phase, and it is connected with a second phase called *characterisation phase,* which corresponds to site maturation and testing.

Those criteria should comprise both technique and socio-economic criteria, and they should answer several questions such as where, how much quantity, and which conditions. All of these criteria and questions will contribute to solve and select the most suitable emplacement for storing CO_2 [10].

There are different examples and analogues that can be useful for the definition of criteria. Analogues can be natural (releases and resources) and industrial.

Assess or characterisation task is related to three major ways to explore the sub-surface: outcrops, geophysics and wells.

To decrease the inherent risk of exploration, it is necessary to consider all of the three sub-phases:

- Outcrops exploration provides samples of seal and storage formation, to evaluate some properties such as hydrogeology and geomechanical properties (permeability, porosity, etc.)

- Geophysics survey will provide and describe geological structures, and in some cases hydrogeology parameter (i.e., total dissolved solid, TDS).

- Few techniques may be used if the structure should be 1,000 m deep: seismic reflection is the most important technology (Figure 10), but other technologies such as magneto-telluric or gravimetric may provide relevant information regarding to the geological structure and resistivity of the original fluid.

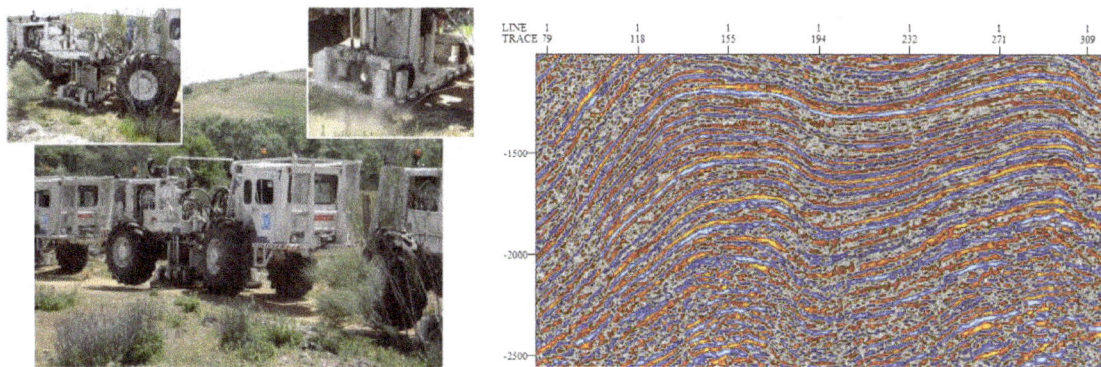

Figure 10. Seismic survey based on vibroseis. Example of a seismic profile.

- Wells will provide real information of the storage and caprock formation in sub-surface conditions. Test will provide information about geomechanical, hydrogeological properties and it may be possible to test interaction between the rock and CO_2.

7. Monitoring techniques

Considering the injection phase, control of the behaviour of injected CO_2 is one of the most important tasks. For instance, the control and monitoring strategy must:

- Demonstrate that the injected CO_2 is stored in the selected reservoir and therefore must be a guarantee that the company responsible for fulfilling its commitment to reducing emissions.

- Check that no intrusion occurs in other exploitable aquifers and water resources.

- Check for surface environmental effects occur, and therefore, you must provide the affected population security and peace on the operations of injection.

The monitoring strategy should not be limited to the operational and post-operational periods but has an important role during the pre-operational stage by conducting the baseline of the injection site [11]. This baseline defines the set of physical, geochemical and biological processes operating in the storage area before any activity injection. The baseline is critical because especially in the early stages of injection, the changes are not evident, both in depth and on the surface, and comparison with the undisturbed condition is needed. The development of the baseline may have added value; for example, building trust and showing the population from the beginning that the project is under control and that any anomaly is detected. Numerous methods have been proposed for monitoring CO_2 in geological repositories. Of these, one can clearly distinguish two types: (a) to detect the evolution of CO_2 injected into deep and (b) for leakage from storage. In the first type, these methods are generally based on geophysical techniques, while in the second type, the range of methods is broader, including geochemical, physical and biological techniques. Therefore, the final selection of the monitoring strategy should take into account the following aspects [12, 13]:

- Efficiency in detecting small changes in behaviour warehouse

- Implementation in large tracts of land

- Reasonable economic cost

Compliance with these requirements will be conditioned by the type of store and its area of influence. Clearly, monitoring techniques will be very different in stores on-shore and off-shore, and within a storage type, geological, hydrological and even ecological characteristics will favour the implementation of a methodology or other.

The monitoring deployment is based on the following aspects: (a) characterisation of the area, (b) establishing a base line CO_2, (c) establishment of potential areas of migration and release of CO_2 (and other gases) and (d) validation and development of techniques for monitoring CO_2.

Emission	Leakage	Description/Comments	Phases of the CO_2 Storage
Leak paths through wells and boreholes	Operating or abandoned wells	It is important to identify all abandoned at the site (or close to it) wells These wells can become the main roads of leakage	· Characterisation · Injection · Post-injection
	Blowouts (uncontrolled emissions from injection wells)	May cause leakage of high flows in short periods of time. It is considered that it is unlikely incidents if safety standards are met during drilling	· Injection
	Future removal of reservoirs of CO_2 can be a problem in the reservoirs of coal deposits	Can be a problem in the reservoirs of coal deposits	· Post-injection
Leak paths and natural migrations through faults and fractures	Through faults and fractures	May origin leak high flows Proper site characterisation can reduce the risk of leakage	· Characterisation · Injection · Post-injection
	By dissolving CO_2 in a fluid and subsequent transport natural circulation of fluid	Proper site characterisation (evaluation of hydrogeology) can reduce the risk of leakside of gas	· Characterisation · Injection · Post-injection
	Through the pore system in low permeability rocks when the capillary inlet pressure is exceeded or if the CO_2 has been in solution	Proper site characterisation can reduce the risk of leaks An exhaustive control of the injection pressure is needed	· Characterisation · Injection
	By a stroke if the reservoir	Proper site characterisation overflows can reduce the risk of leakage	· Injection · Post-injection
Another type of leak	Methane release could occur as a result of the displacement of CH_4 by CO_2	It happens in depleted oil and gas	· Characterisation · Injection · Post-injection

Table 3. Possible types of leakage of CO_2 in a geological storage [13, 14, 15].

8. CO_2 uses

8.1. Current uses of CO_2

Nowadays different applications are known that can be used for demonstrating that CO_2 is a useful, versatile and safe product. Figure 11 illustrates most of the current and potential uses of CO_2.

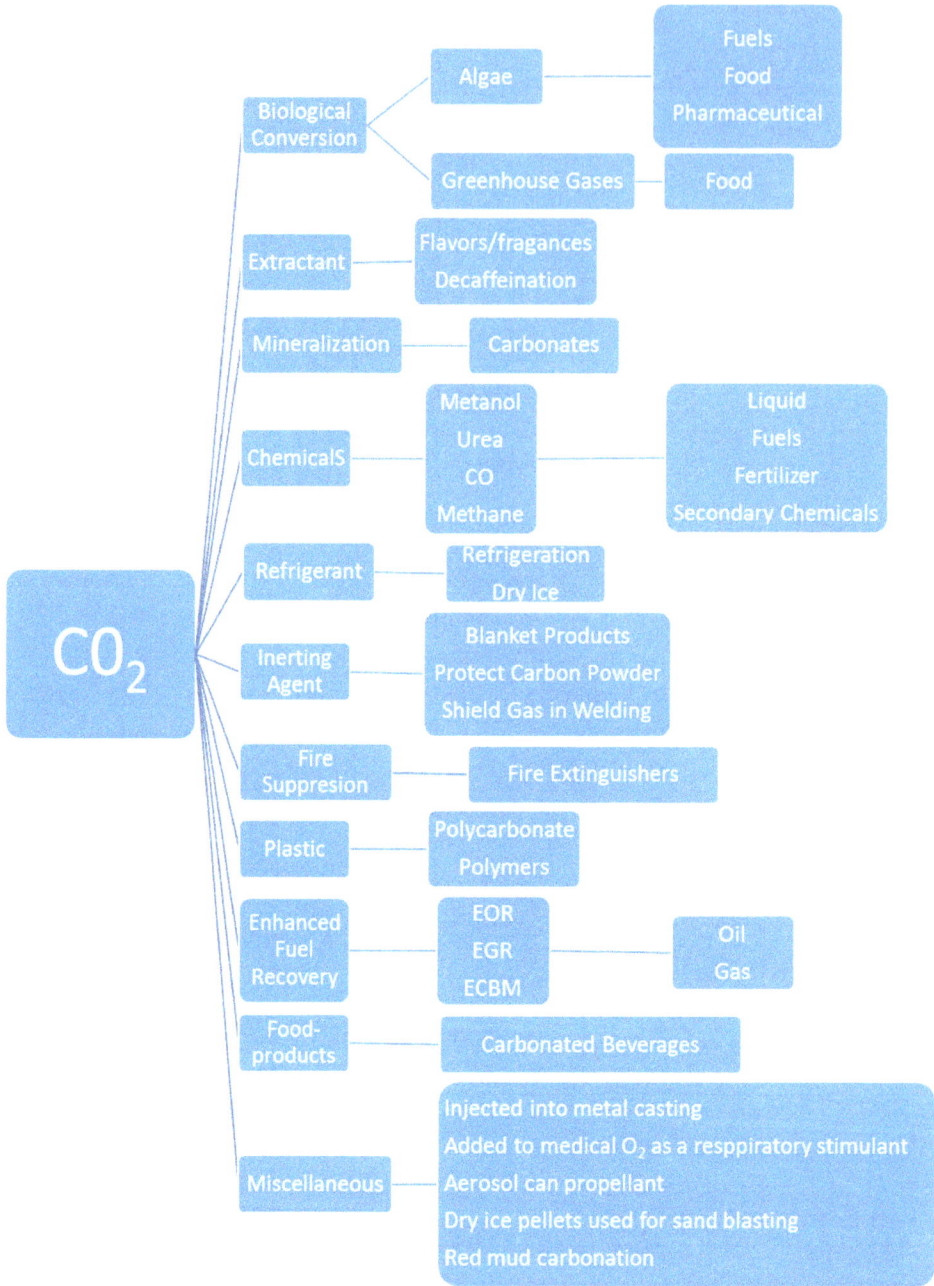

Figure 11. CO_2 uses. Different pathways for utilisation CO_2.

There are many classifications that can be made about the use or valuation of large-scale CO_2 and including the three categories proposed by Vega [16] for type of uses, which also is used by the PTE-CO2, 2013 (Technology Platform Spanish CO_2). To wit:

1. Direct or technology use: use of CO_2 with different technologies and market applications such as use for oil recovery, for dry cleaning, waste carbonation, food, water treatment or extraction with supercritical CO_2 compounds, including others.

2. Improved biological use: CO_2 fixation in biomass by growing microalgae and carbonic fertigation.

3. Chemical use: artificial photosynthesis and chemical conversion to high added value products and fuels.

8.2. Direct use of CO_2

8.2.1. Enhanced oil recovery (CO_2-EOR)

As much as two-thirds of conventional crude oil discovered in U.S. fields remain unproduced, left behind because of the physics of fluid flow. In addition, hydrocarbons in unconventional rocks or that have unconventional characteristics (such as oil in fractured shales, kerogen in oil shale or bitumen in tar sands) constitute an enormous potential domestic supply of energy.

Carbon dioxide is used in oil wells for oil extraction and to maintain pressure within a formation.

There are many methods for EOR and each has differences that make it more useful based on specific reservoir challenges and other parameters. Choosing the right method by screening the reservoir and fluid properties can ultimately reduce risk by eliminating inefficiencies.

CO_2 EOR is an 'EOR' technology that injects CO_2 into an underground geologic zone (oil/ hydrocarbon containing 'reservoir') that contains hydrocarbons for the purpose of producing the oil. The CO_2 is produced along with the oil and then recovered and re-injected to recover more oil.

When the maximum amount of oil is recovered from the reservoir, the CO_2 is then 'sequestered' in the underground geologic zone that formerly contained the oil and the well is shut-in, permanently sequestering the CO_2.

CO_2 injection is a technology successfully used from more than 50 years. The first patent for CO_2-EOR appeared in 1952 and in 1964 began field trials. In the first commercial project of CO_2-EOR in Texas, in 1972 (SACROC project), CO_2 was supplied from a gas plant, where the CO_2 was eliminated in the production of ammonia At present the CO_2 is sent from geological formations (natural) from Bravo Dome in Colorado, and Mc Elmo Dome in New Mexico.

Nowadays, two techniques are largely used for CO_2-EOR:

- **Miscible water-alternating-gas (WAG) process.** Injection alternates between gas (usually natural gas or CO_2) and water; the miscible gas and oil form one phase. The WAG cycles improve sweep efficiency by increasing viscosity of the combined flood front (Figure 12).

- **Cyclic gas injection.** Most gas-injection EOR projects today use CO_2 as the injected gas. When CO_2 is pumped into an oil well, it is partially dissolved into the oil, rendering it less

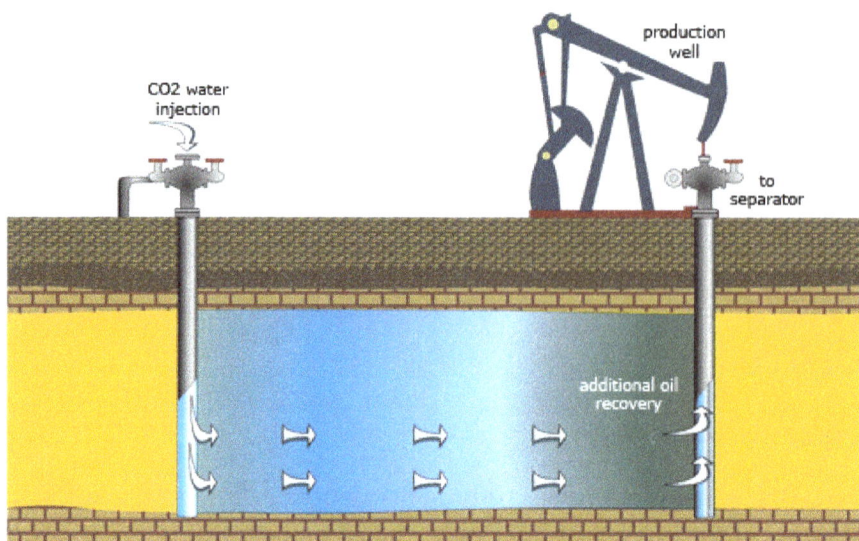

Figure 12. CO_2-EOR operation diagram. CO_2 injection into reservoir to 'flood'. Diagram courtesy of Dakota Gasification Company.

viscous, allowing the oil to be extracted more easily from the bedrock. The CO_2 used to increase oil recovery can be naturally occurring, or an effective means of sequestering an industrial by-product. In this case, carbon dioxide, under pressure, is injected between oil wells to freeing the stranded oil. CO_2 is a superior agent in recovering stranded oil as the CO_2 naturally reduces the surface tension that traps the liquid oil to in the oil reservoir. When the oil is recovered from the production well, CO_2 is also produced, but is easily separated from the crude oil because the CO_2 reverts back to its gaseous state when the pressure is removed.

8.2.2. Fire suppression

Some fire extinguishers use CO_2 because it is denser than air. Carbon dioxide can blanket a fire, because of its heaviness. It prevents oxygen from getting to the fire and as a result, the burning material is deprived of the oxygen it needs to continue burning.

8.2.3. Supercritical CO_2 uses

When CO_2 is at suitable temperature and pressure above the critical point (Figure 13), it is called supercritical CO_2.

This state emphasises its capacity to dissolve chemicals and natural substances of similar way as do different organic solvents such as hexane, acetone or dichloromethane. Therefore, the first applications focused on **the extraction of natural substances** as an alternative to using organic solvents. Thus, **removal of caffeine** (coffee or tea) with supercritical CO_2 is the most mature application at industrial level and is also used in the **extraction of hops or cocoa fat.**

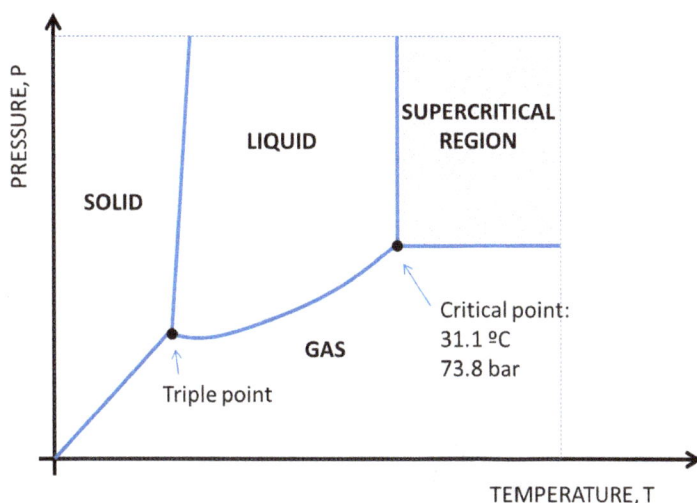

Figure 13. CO_2 phases diagram.

The dry cleaning with CO_2 is one of the most popular applications of supercritical fluids in the textile sector. This method is characterised by removing stains from the fabrics and garments where no harmful organic solvents for the average ambient, such as perchlorethylene (PER), common in conventional dry cleaning processes are used and without causing discoloration or shrinkage and without leaving odour.

One of the main advantages of supercritical CO_2 is that its solubility can easily be controlled suitably adjusting the pressure and temperature, allowing **fractionate mixtures** where all components are soluble.

Supercritical CO_2 extraction coupled with a fractional separation technique is used by producers of **flavours and fragrances** to separate and purify volatile flavour and fragrance concentrates. Like any solvent, supercritical CO_2, it allows processing chemicals by precipitation or recrystallisation, obtaining particles of controlled size and shape, without excessive fines without thermal stresses and controlling the shape of a polymorphic substance.

It is, therefore, a cutting-edge technology with great potential, because it is a new way to obtain natural products; it allows the adaptation of new high quality products with appropriate value to consumer habits; enables the development of new non-polluting processes and initiate the development of a tertiary sector led to the new technology.

8.2.4. Food and beverages

Liquid or solid CO_2 is used for quick freezing, surface freezing, chilling and refrigeration in the transport of foods. In cryogenic tunnel and spiral freezers, high pressure liquid CO_2 is injected through nozzles that convert it to a mixture of CO_2 gas and dry ice 'snow' that covers the surface of the food product. As it sublimates (goes directly from solid to gas states), refrigeration is transferred to the product.

Carbon dioxide gas is used to carbonate soft drinks, beers and wine and to prevent fungal and bacterial growth. CO_2 has an inhibitory effect on bacterial growth, especially those that cause discoloration and odours.

CO_2 has an inhibitory effect on bacterial growth, especially those that cause discoloration and odours (Figure 14).

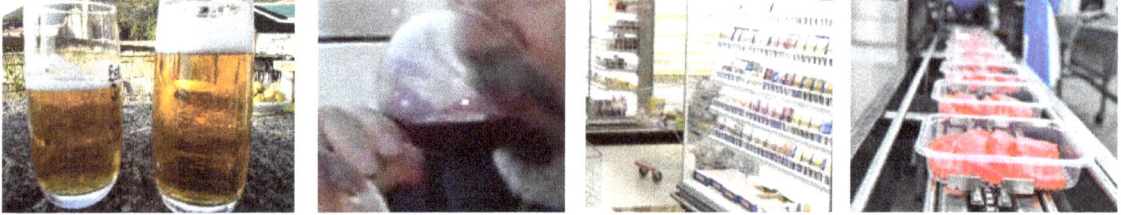

Figure 14. CO_2 applications in food.

It is used as an inert 'blanket', as a product-dispensing propellant and an extraction agent. It can also be used to displace air during canning.

Cold sterilisation can be carried out with a mixture of 90% carbon dioxide and 10% ethylene oxide, the carbon dioxide has a stabilising effect on the ethylene oxide and reduces the risk of explosion.

8.2.5. Water treatment

Carbon dioxide can change the pH of water because of its slightly dissolution in water to form carbonic acid, H_2CO_3 (a weak acid), according to Equation 9:

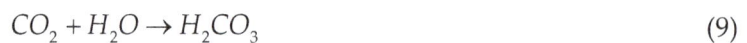

$$CO_2 + H_2O \rightarrow H_2CO_3 \qquad (9)$$

Carbonic acid reacts slightly and reversibly in water to form a hydronium cation H_3O^+, and the bicarbonate ion HCO_3^-, according to Equation 10:

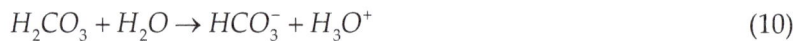

$$H_2CO_3 + H_2O \rightarrow HCO_3^- + H_3O^+ \qquad (10)$$

This chemical behaviour explains why water, which normally has a neutral pH of 7 has an acidic pH of approximately 5.5 when it has been exposed to air.

At the moment, CO_2 technology is widely introduced in treatments such as sewage water, industrial water or drinking water remineralisation.

The increased requirements of drinking water in large cities becomes necessary to use sources of very soft water and because of its low salinity and pH are very aggressive and can bring on corrosion phenomena in the pipes of the pipeline, with the appearance of colour and turbidity when these pipes are made of iron, and by undermining these ones made with cement fibre by dissolving the calcium carbonate ($CaCO_3$), because of excessive aggressive CO_2.

The introduction of carbon dioxide in the pipes regulates a state of equilibrium between dissolved bicarbonates, calcium carbonate inlaid and the CO_2 added.

Therefore, for the treatment of soft or aggressive waters, the use of CO_2 in combination with lime or calcium hydroxide is advisable to increase water hardness. This process is called **remineralisation** and is meaningful in water treatment plants, because soft water is indigestible.

The use of CO_2 in wastewater neutralisation, Figure 15, offers great advantages in the operation and the environment by preventing other chemicals:

- Better working conditions. Eliminate the risk of burns, toxic fumes and other injuries from handling mineral acids

- Safe neutralisation. Avoiding risk of over-acidification with strong acids

- Low initial investment. Simple equipment, insurance and small dimensions

- Automated process. Automation avoids the handling of corrosive acids in the plant, pH control is automatic

- Economy

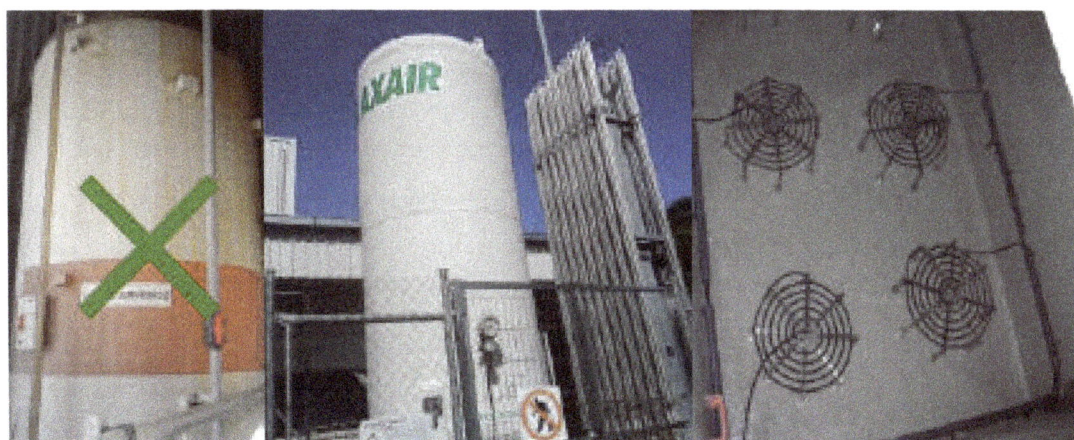

Figure 15. Dosing system for sewage.

8.2.6. Carbonate mineralisation

The alkaline waste management presents significant problems, mainly because of its volume and its geochemical properties that do not allow disposing in conventional landfills. Therefore, the accelerated carbonation of this waste is another technological uses of CO_2.

Carbonate mineralisation refers to the conversion of CO_2 to solid inorganic carbonates. Naturally occurring alkaline and alkaline-earth oxides react chemically with CO_2 to produce minerals, such as calcium carbonate ($CaCO_3$) and magnesium carbonate ($MgCO_3$). These minerals are highly stable and can be used in construction or disposed of without concern that the CO_2 they contain will release into the atmosphere. One problem is that these reactions tend to be slow, and unless the reactions are carried out in situ, there is a large volume of rocks to move. Carbonates can also be used as filler materials in paper and plastic products.

8.2.7. Biological utilisation

Green plants convert carbon dioxide and water into food compounds, such as glucose and oxygen. This process is called photosynthesis (Equation 11).

$$CO_2 + 6H_2O \longrightarrow C_6H_{12}O_6 + 6O_2 \qquad (11)$$

The reaction of photosynthesis is as follows: Biological applications are based primarily on the use of CO_2 as food for plant growth. In a similar way as the plants take advantage of sunlight and CO_2 for biomass, or other products, 'imitating' nature, improving its results. Therefore, this technology is also known as biomimetic transformation.

There are two main ways in the biological utilisation process: greenhouses carbonic fertilisation and growth of microalgae.

8.2.7.1. Greenhouses carbonic fertilisation

CO_2 is found naturally in the atmosphere and, therefore, in the **greenhouse** environment. It is essential for plant growth, since it represents the carbon source for organic compounds they need, in short, for compounds that constitute their biomass (leaves, stems, fruits, etc.).

CO_2 is not the only factor involved in photosynthesis, so that for its use, other factors must be at levels that do not limit the process. Light, temperature, amount of available nutrients and the relative humidity are other environmental factors affecting photosynthetic activity.

During photosynthesis, plants capture light energy and CO_2 through the leaves, and water and nutrients through the roots. Thanks to these elements and chlorophyll leaves, plants get synthesise sugars and various organic compounds required for their development. Photosynthesis is responsible for plant growth. Therefore, favouring photosynthesis we managed to promote the development of the plants and agriculture in our case.

Yields of plant products grown in **greenhouses** can increase by 20% by enriching the air inside the greenhouse with carbon dioxide. The target level for enrichment is typically a carbon dioxide concentration of 800 ppm – or about two-and-a-half times the level present in the atmosphere (Figure 16).

In the CENIT SOST-CO_2 project that includes the entire life cycle of CO_2, researching technology uses as chemical and biological uses, the following results were confirmed, among others [18]:

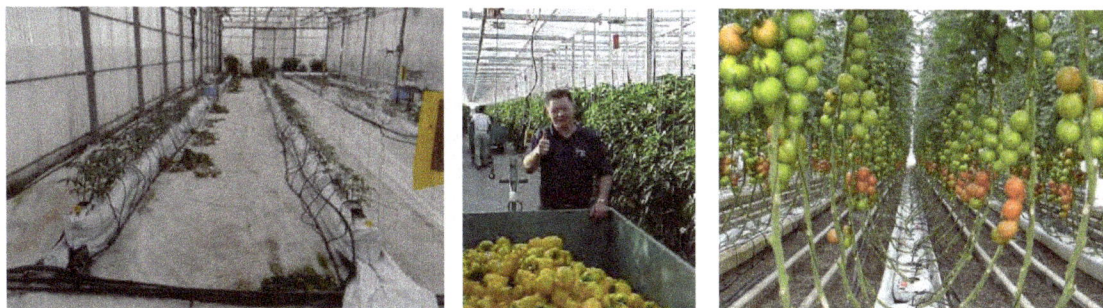

Figure 16. Carbon fertilisation in hydroponics culture greenhouses.

- Rating combustion gases from combined cycle plants to use in vegetable crops in greenhouses, in applications in irrigation pipes to prevent clogging and to balance the pH in nutrient solutions.

- With regard to the quality of gas from CCGT, it can be recommended for direct use in greenhouses or other agricultural uses.

- The carbonic fertilisation allows early crop production along with a greater amount of product with better quality.

8.2.7.2. Growth of microalgae

Microalgae are photosynthetic microorganisms that can grow in diverse areas mainly in water media where the forced culture can be carried out in diverse type of reactors in concordance with its design and operation. The advantage of this process is that microalgae are a microorganism with a high production rate (some species are able to duplicate their biomass in 24 hours), and therefore with increased demand for CO_2 conventional terrestrial plants.

The investigation of microalgae culture for different purposes began in the middle of last century, when the United States launched the 'Aquatic Species Program'. At that time, the research focused on the possibility of obtaining biofuels from microalgae: mainly methane and hydrogen, but after the oil crisis in the 1970s the biodiesel was also considered.

Biofixation of CO_2 by microalgae, especially as an option for the utilisation of flue gases from power plants, has been the subject of extensive investigations in the United States, Japan and Europe (IEA-GHG Biofixation Network). However, none of the related projects have demonstrated the feasibility of the concept at a pre-industrial level. What is more, CO_2 fixation efficiency is quite low because of the photobioreactors used in those pilot plants (raceway or open-ponds) (Figure 17).

The current production of microalgae is mainly focused around a few species, such as *Spirulina*, *Chlorella*, *Dunaliella* or *Haematococcus* for nutritional purposes (for humans) and animal feed (especially aquaculture). Other sectors, such as cosmetics, effluent treatment and bioenergy, have shown interest, incorporating these or other species of microalgae and cyanobacteria into commercial products. Currently, 95% of the production of microalgae is based on open systems

Figure 17. Microalgae culture in open system (raceway) and close photobioreactor (Almeria University and Palmerillas Research Center).

(raceways or circular open ponds). These systems have a low rate of CO_2 fixation and it is estimated to be around 20–50% of the injected gas is effectively set by microalgae [17].

8.2.8. Use of CO_2 in chemicals

Carbon dioxide gas is used to make urea (used as a fertiliser and in automobile systems and medicine), methanol, inorganic and organic carbonates, polyurethanes and sodium salicylate. Carbon dioxide is combined with epoxides to create plastics and polymers.

Corn-to-ethanol plants have been the most rapidly growing source of feed gas for CO_2 recovery.

8.2.8.1. Artificial photosynthesis

Because CO_2 is a practically inert molecule, artificial photosynthesis of CO_2 involves the use of large amounts of energy so it must use a clean source of energy (such as solar radiation).Therefore, the use of catalytic agent to facilitate the process allowing even take place at ambient temperature and pressure is necessary. In this case, it is also called as photocatalysis or photoreduction.

In photocatalysis two processes occur: CO_2 reduction and oxidation of other compounds. Early works on the photocatalytic reduction of CO_2 in aqueous solution were published between 1978 and 1979 ([19, 20]), and later numerous investigators have studied the mechanism and efficiency of the process using different catalysts (oxides of titanium, zinc and cadmium, cadmium sulphide, silicon carbide), and reducing (water, amines, alcohols) and R light sources (lamps xenon, mercury, halogen). Thus, it has been shown that by using specific semiconductors and reducing agents, can be obtained a great variety of products (methane, methanol, formaldehyde, formic acid, ethanol, ethane, etc.).

Along with thermodynamics, catalysis is one of the core technologies for an economically interesting use of CO_2 as feedstock in chemical processes. This is one of the areas most sophisticated and complex of modern chemical research. It is one of the major challenges for the scientific and technological developments related to the fields of energy and catalysis, as was highlighted in the report to officiate Sciences US Department basic Energy: more than 85% of all products are produced using chemical catalysis [21].

Photocatalysis involve the production of reactions because of the incidence of light on a semiconductor material. Unlike metals, these materials have a forbidden energy band, which extends from the top of the so-called valence band to the bottom of the conduction band (Figure 18).

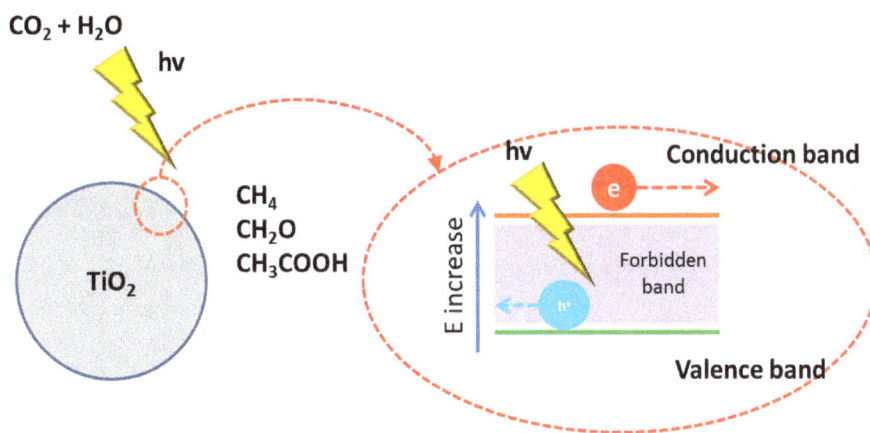

Figure 18. Diagram of behaviour of a semiconductor, TiO_2, in light presence and participation in the photocatalytic CO_2 reduction organic products.

The main disadvantage in these cases remains in the low process efficiency.

In general, the process of photocatalytic reduction of CO_2 requires a milder conditions and lower energy consumption than chemical reduction [22].

8.3. Chemical conversion

Large quantities are used as a raw material in the chemical process industry, especially for urea across CO_2 reaction with NH_3 and later dehydration of the formed carbamate. Urea is the product most used as agricultural fertiliser. It is used in feed for ruminants, as carbon cellulose explosives stabiliser in the manufacture of resins and also for thermosetting plastic products, among others.

Methanol production, where CO is added as additive, is very a well-known reaction. The production is carried out in two steps. The first step is to convert the feedstock natural gas into a synthesis gas stream consisting of CO, CO_2, H_2O and hydrogen. This is usually accomplished by the catalytic reforming of feed gas and steam. The second step is the catalytic synthesis of methanol from the synthesis gas. If an external source of CO_2 is available, the excess hydrogen can be consumed and converted to additional methanol.

CO_2 is also used, to make inorganic and organic carbonates, carboxylic acids, polyurethanes and sodium salicylate. Carbon dioxide is combined with epoxides to create plastics and polymers (Figure 19).

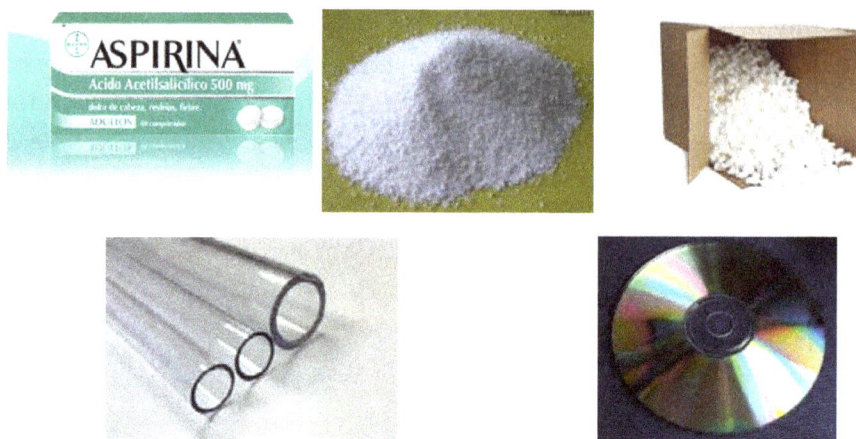

Figure 19. Different products made with CO_2 derivatives.

9. New ways for CO_2 uses

In general, the area of CO_2 utilisation for carbon storage is relatively new and less well known compared to other storage approaches, such as geologic storage. Thus, more exploratory technological investigations are needed to discover new applications and new reactions.

Many challenges exist for achieving successful CO_2 utilisation, including the development of technologies capable of economically fixing CO_2 in stable products for indirect storage.

Significant innovation and technical progress are being made across a number of utilisation technologies. The **electrochemical reduction** could be really attractive because it is an excellent way for renewable energy storage.

9.1. Power to gas technology (P2G)

In the 3rd Carbon Dioxide Utilisation Summit, October 2014 in Bremen, Germany, ETOGAS GmbH presented its turn-Key plan and technology Power-to-Gas for SNG through electrolysis processes [18].

This technology uses CO_2 as a feed gas for the production of carbon products with Etogas methanation plant (Figure 20), which are reactor systems for conversion of H_2 and CO_2 to methane (synthetic natural gas). The produced gas is DVGW- and DIN-compliant synthetic natural gas and can be used directly, e.g., as a fuel for a CNG vehicle.

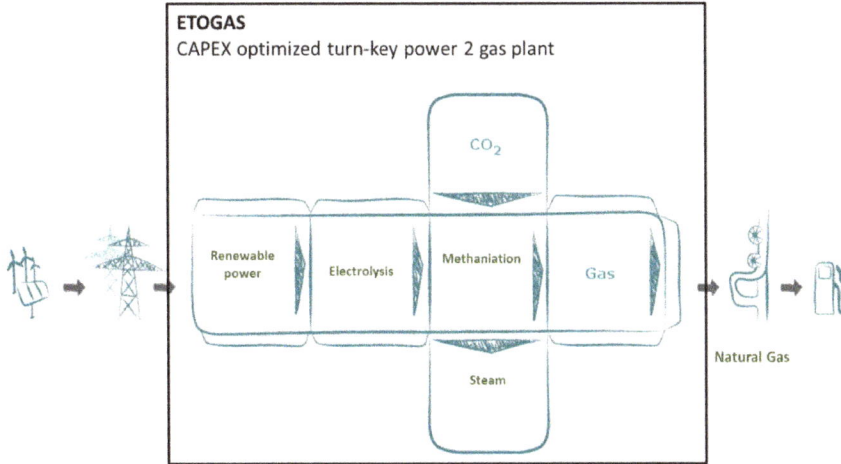

Figure 20. SNG schematic process. Source: ETOGAS Project.

9.2. Electrochemical CO_2 utilisation

According to DNV GL, electrochemical CO_2 utilisation presents some advantages as follows:

Production de-coupled from the sun (flexibility in renewable energy source); land use is minimised and no limitation with respect to geography; no competition with food (corn, sugar); flexibility in end fuel – ethanol, butanol or diesel (depending on the organism used); flexibility in electrochemical process (matching to supply/demand of renewable energy); and significant net reduction in CO_2 emission (Figure 21).

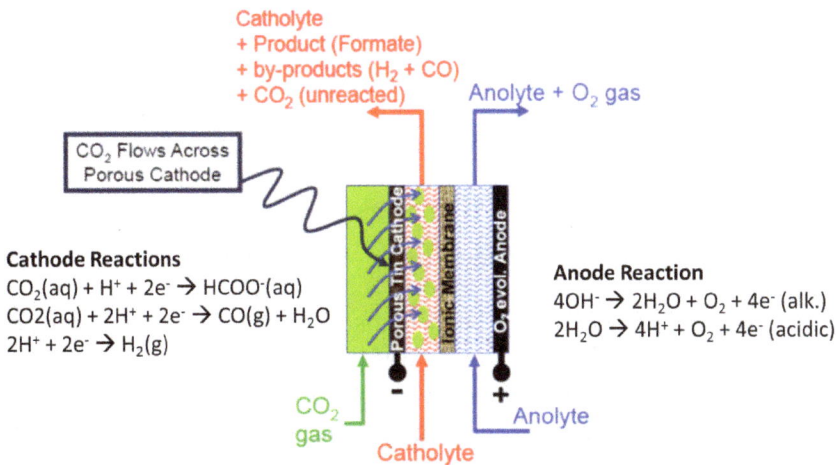

Figure 21. Electrochemical production of formic acid (HCOOH) and CO. Source: third Carbon Dioxide Utilisation Summit. DNV GL.

9.3. Polymers production

Bayer MaterialScience (Germany) in the Project "Dream Production" combines part of waste streams of coal-fired power plants, CO_2, with the production of polymers. The target is the design and development of a technical process able to produce CO_2-based polyether polycarbonate polyols on a large scale. The first step was to convert the CO_2 in new polyols, and these polyols showed similar properties such as products already on the market and can be processed in conventional plans as well (Figure 22).

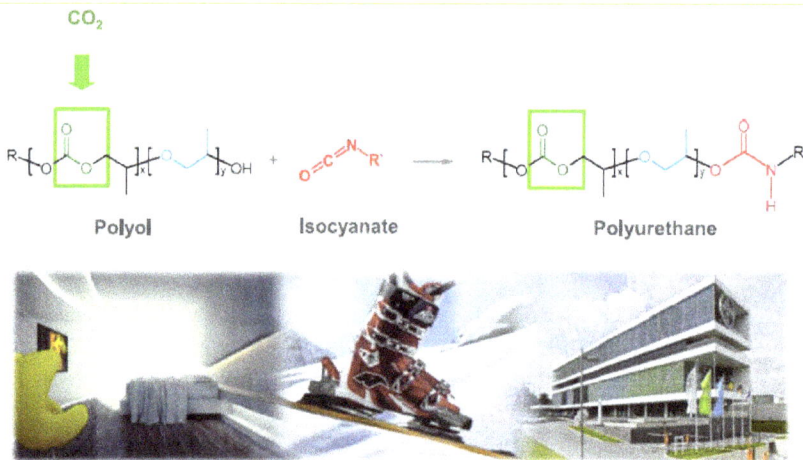

Figure 22. Target product polyurethanes – All rounder among plastics. Source: 3rd Carbon Dioxide Utilisation Summit. Courtesy: Bayer.

The CO_2 thus acts as a substitute for the petroleum production of plastics. Polyurethanes are used to produce a wide range of everyday applications. When they are used for the insulation of buildings, the polyurethane saves about 80% more energy than it consumes during production. Light weight polymers are used in the automotive industry, upholstered furniture and mattress manufacturing.

9.4. Macrofouling control in industrial facilities

In the past years, several projects have been focused in the **direct use of flue gases** from Combined Cycle Power Plants for developing different applications. In this way, the project CENIT SOST-CO2 has demonstrated the use of flue gases from CCPP in a direct way to control the pH in the cooling water systems with refrigeration tower and Iberdrola has developed an application for power plants.

Another application for the future will be "CO_2 for Zebra Mussel Control". A project developed by Iberdrola and the University of Salamanca shows that carbonic acidification just in the moment when the larva of zebra mussel are in the adequate phase (pediveliger) causes a much greater lethality than inorganic acids because of the synergistic effect of the lethal hypercapnia by physiological changes in cell metabolism of the larvae. (*CDTI Project: Seguimiento de la incidencia del mejillón cebra (Dreissena polymorpha) en el Ciclo Combinado de Castejón 2009-2011. Iberdrola – Universidad de Salamanca*).

The project LIFE13 ENV/ES/000426. CO2FORMARE [23], "Use of CO_2 as a substitute of chlorine-based chemicals used in O&M Industrial processes for macrofouling remediation", led by Iberdrola Generación, seeks to demonstrate the viability of using CO_2 from combustion gases to control macrofouling (*fouling caused by larger organisms, such as oysters, mussels, clams and barnacles*) in a thermal power plant (Castellon CCPP), cooled by sea water. First estimates indicate that a 400 MW CCPP (Figure 23) may be necessary to use up to 50,000 t CO_2 yr^{-1}, [23].

Figure 23. Castellon Power Plant. Courtesy: Iberdrola.

10. Others

The Carbon Storage Program of the NETL (National Energy Technology Laboratory) of US Department of Energy supports four main CO2 utilisation research areas: **cement**, **polycarbonate plastic**, **mineralisation** and **enhanced hydrocarbon recovery.** Several projects on active CO_2 utilisation focused in these areas receive Department of Energy (DOE) funds that aim to obtain the goals for the Carbon Storage Program.

Author details

Bernardo Llamas[1,2*], Benito Navarrete[3], Fernando Vega[2], Elías Rodriguez[4,5],
Luis F. Mazadiego[1], Ángel Cámara[1] and Pedro Otero[6]

*Address all correspondence to: bernardo.llamas@upm.es

1 Escuela de Ingenieros de Minas y Energía, Universidad Politécnica de Madrid (UPM),
Madrid, Spain

2 INERGYCLEAN Technology, Almería, Spain

3 Escuela Técnica Superior de Ingeniería de Sevilla, Sevilla, Spain

4 IBERDROLA Generación, Madrid, Spain

5 Westec Environmental Solutions, llc, Chicago, USA

6 Es.CO_2. Centro de Desarrollo de Tecnologías de Captura de CO_2, CIUDEN, León, France

Authors, Elías Rodriguez and Pedro Otero, retired from their affiliations.

References

[1] http://www.esrl.noaa.gov/gmd/ccgg/trends/ [Accessed on 1 February 2016].

[2] Intergovernmental Panel on Climate Change: Climate Change 2013: The Physical Science *Basis*. Cambridge University Press. UK, 2013.

[3] International Energy Agency. (ed.): *Key World Energy Statistics 2015*. International Energy Agency publication. 2013, 82 pp.

[4] Global CCS Institute: *The Global Status of CCS*. Global CCS Institute. 2015. 24 pp.

[5] E. Rubin, H. Mantripragada, A. Marks, P. Versteeg, J. Kitchin: *The outlook for improved carbon capture technology*. Progress in Energy and Combustion Science, 38(5). 2012.

[6] L. Kohl, R.B. Nielsen: *Gas purification*. ELSEVIER. 1997.

[7] Toftegaard, Maja Bøg et al. *Oxy-fuel combustion of solid fuels*. Progress in Energy and Combustion Science. 2010, 36(5). 581-622.

[8] P. Noothout, F. Wiersma, O. Hurtado, P. Roelofsen, D. Macdonald: CO_2 pipeline infrastructure, report 2013/18. IEA GHG. 2014. United Kingdom.

[9] O. Skovholt: *CO_2 Transportation System*. Energy Conversion & Management. 1993. Vol. 34, 9–11, 1095–1103.

[10] B. Llamas, P. Cienfuegos: *Multicriteria Decision Methodology to Select Suitable Areas for Storing CO₂*. Energy & Environment. 2012. Vol. 23, 2–3, 249–264.

[11] R.W. Klusman: Baseline Studies of Surface Gas Exchange and Soil-Gas Composition in Preparation for CO_2 Sequestration Research: Teapot Dome, Wyoming. The American Association of Petroleum Geologists. AAPG Bulletin. 2005. 89, 981–1003.

[12] M. Ortega, M.A. Rincones, J. Elío, J. Gutiérrez, B. Nisi, L.F. Mazadiego, L. Iglesias, O. Vaselli, F. Grandia, R. de la Vega, B. Llamas: *Gas Monitoring Methodology and Application to CCS Projects as Defined by Atmospheric and Remote Sensing Survey in the Natural Analogue of Campo de Calatrava*. Global Nest Journal. 2014. 16(2), 269–279.

[13] J. Elío, M. Ortega, B. Nisi, L.F. Mazadiego, O. Vaselli, J. Caballero, L. Quindós-Poncela, C. Sainz-Fernández, J. Pous: *Evaluation of the Applicability of Four Different Radon Measurement Techniques for Monitoring CO_2 Storage Sites*. International Journal of Greenhouse Gas Control. 2015a. 41, 1–10.

[14] Llamas, B.; Mazadiego, L.F.; Elío, J.; Ortega, M.; Grandia, F.; Rincones, M.A. System-atic *Approach for the Selection of Monitoring Technologies in CO_2 Geological Storage Projects*. Application of *Multicriteria Decision Ma*king. Global Nest Journal. 2014. (16-1), pp. 36 - 42.

[15] R.W. Klusman: Comparison of Surface and Near-Surface Geochemical Methods for Detection of Gas Microseepage from Carbon Dioxide Sequestration. International Journal of Greenhouse Gas Control. 2011. 5, 1369–1392.

[16] L. Vega: El CO_2 como recurso. De la naturaleza a los usos industriales. Gas Natural Fenosa Foundation. 2010.

[17] Technology Platform Spanish CO₂.Usos del CO_2: un camino hacia la sostenibilidad. 2013.

[18] http://www.cenit-sostco2.es/ [Accessed on 1 December 2015].

[19] M. Halmann. Photoelectrochemical reduction of aqueous carbon dioxide on p-type gallium phosphide in liquid junction solar cells. Nature. 1978, 275(5676): 115-116.

[20] Inoue, T., Fujishima, A., Konishi, S. and Honda, K. Photoelectrocatalytic Reduction of Carbon Dioxide in Aqueous Suspensions of Semiconductor Powders. Nature. 1979, 277: 637.

[21] US Department of Energy, DOE: *Basic research needs, Catalysis for Energy*. 2007. Basic Energy Sciences Workshop.

[22] Kohno, Y., Hayashi, H., Takenaka, S., Tanaka, T., Funabiki, T. and Yoshida, S. *Photo-enhanced reduction of carbon dioxide with hydrogen over Rh/TiO_2*. Journal of Photochemistry and Photobiology. 1999. A 126, 117-123.

[23] http://www.etogas.com/ [Accessed on 1 December 2015].

7

Native Forest and Climate Change — The Role of the Subtropical Forest, Potentials, and Threats

Silvina M. Manrique and Judith Franco

Additional information is available at the end of the chapter

Abstract

The subtropical rainforest of Argentina, called Yungas, has been subjected to rapid deforestation and degradation processes in recent years, especially in the lower district: "Pedemontana Jungle" (PJ; ≤900 m.a.s.l.). In Salta, in the north of the country, the rate of deforestation is around three times higher than the world average. The disappearance of PJ significantly limits the area of contact between Yungas and Chaco forest, which could have important consequences for natural and cultural biodiversity in the region (the largest number of aboriginal ethnic groups live here, most of which depend on native forest for their existence and identity). In addition, the loss and degradation of forests is the second largest sector of greenhouse gas (GHG) emissions to the atmosphere (about 18%), affecting the world climate. We present a synthesis of different studies developed in PJ forests, observing its role as reservoirs of carbon and discussing issues that could influence the total capacity of carbon sequestration of the same. This will contribute to build the reliable database on the sequestration potential, which will facilitate standardization of units, reduction of uncertainties, and contribution to a more efficient strategy to limit the GHG emission to the environment, providing some learning and useful recommendations.

Keywords: biomass, carbon sequestration, edge effect, fragmentation, native forest

1. Introduction

1.1. Deforestation, fragmentation, and climate change

According to recent studies, the forests covering about 30% of the earth's surface [1] contain 80% terrestrial biomass and provide habitat for about half of the world's known species of plants and animals [2]. Forests provide a wide range of ecological, economic, and social assets, as well as services such as climate regulation through the storage of carbon in complex physical, chemical, and biological processes [3–5]. Despite a wide recognition of the importance of native forests, recent data show that the loss of forest cover over the planet (deforestation) in 2000–2012 was 2.3 million km^2, while the gain (grown or planted) was 0.8 million km^2 [6]. Conversely, Keenan et al. reported a rate of 0.08% of forest loss in 2010–2015, while farmland continued expanding in 70% of the countries [1].

Native forests have been affected in terms of not only the total amount of existing surface (deforestation) but also the quality of the remaining fragments (degradation) [5, 7], therefore the biomass availability and its derived flow, which means a source of ecosystem goods and services, has been doubly modified. Several of them, such as soil protection, gas and climate regulation, water regulation, nutrient cycling, providing habitat and refuge, food production, raw materials and genetic resources, the provision of medicinal and ornamental resources, and others related to culture (recreation, aesthetics, and spirituality), are associated with biomass existence and generation [2, 4, 5]. Similarly, there are an increasing number of studies showing the interrelationship between the aboveground and subterranean processes, and particularly among the aboveground biomass (AGB) and soil, links that determine the abundance of species, coexistence, and succession [8, 9]. Therefore, any changes in the biomass, including degradation – although it is a hardly measureable phenomenon [10] – will affect soil characteristics, which, in turn, will modify reproduction patterns and survival of typical plants in the ecosystem in question and their associated fauna [2, 8, 9].

Deforestation and fragmentation of forests, have been an object of study of the scientific community for many years, but attention to these phenomena has begun to rise from the perspective of their contribution to global warming by greenhouse gas emissions [3, 10–15]. It is recognized that the change in land use (including forest degradation and deforestation) is the second sector of global importance in terms of GHG emissions (so-called LULUCF or land use, land use change and forestry) and is responsible for 20% of total emissions [16]; therefore, it is an important component of human impact on global climate.

Variations in the soil cover are one of the natural and anthropogenic forces that operate on different scales, influencing changes in regional and global climates [3, 13, 16]. Malhi et al. [13] document some interrelations in the Amazon forests, noting that they have a great influence on regional and global climates. They mention that the extraction of water from the soil, through the tree roots up to 10 m depth, and its return to the atmosphere ("perspiration service") is, perhaps, the most important regional ecosystem service. Therefore, the removal of trees through deforestation can become a driver for climate change and a positive feedback for externally forced climate change. In agreement with the other authors, forest loss also results in (i) decreased cloud cover and an increase in insulation; (ii) increase in the reflectance of the earth's surface, approximately offsetting the effect of clouds; (iii) changes in the aerosol loading of the atmosphere from a hyperclean "green ocean" atmosphere to a smoky and dusty continental atmosphere that can modify rainfall patterns; and (iv) changes in the surface roughness (and therefore the wind speed) and a large-scale convergence of atmospheric humidity, which generates precipitation [14, 15]. These large-scale interrelations repeat on lesser scales, although they have not been sufficiently studied.

Deforestation and fragmentation could increase the vulnerability of forests to climate change [2, 3, 17, 18], being two interlinked processes, since deforestation to open up new land for cultivation is concentrated in the periphery of existing forest fragments, reducing them in size and/or making them disappear. Both processes have been recognized as important drivers of biodiversity loss [2, 4, 5, 19–21].

1.2. Climate change in Argentina

In 2015, Argentina presented its Third National Communication (TNC) on Climate Change [22], with an updated GHG inventory as part of the fulfillment of their assumed commitments to the United Nations Framework Convention on Climate Change (UNFCCC). They inform that national emissions in 2012 imply a 0.88% participation in global emissions (429,437 Gg CO_{2eq}). The six sectors surveyed were as follows: (1) energy (43% of total emissions), (2) industrial processes (3.6%), (3) use of solvents and other products (0%), (4) agriculture and livestock (27.8 %), (5) land use change and forestry (LUCF) (21.1%), and (6) waste (6%). Within the LUCF sector – the third most important – the subsector of "forest and other land conversion" contributes 67% of emissions.

Of the total native forests, in 2002 (33 million ha), Yungas occupied 11.2% of the surface (3.7 million ha). A TNC report mentions that the loss of native forests in 2002–2010 was 3.5 million hectares (computed in "conversion of forests and other land") corresponding to the 8% loss of Yungas (280,300 ha), which caused a reduction of 7.5% of the total area. The rest of the removed area corresponded to the Chaco region, whose surface involved 70% of the total forests in the country (larger ecosystem) that year.

In effect, the Intergovernmental Panel on Climate Change (IPCC), in which more than 300 scientists from all over the world participate, warned that, in 2014, 4.3% of global deforestation occurred in Argentina [16]. At a local level, the Secretary of Environment for the Nation published, in the same year, the report "Monitoring of the area of native forest in Argentina," pointing out that between November 2007 (when the National Forest Act was enacted) and the end of 2013, 1.9 million hectares were removed – an average of 1 ha/2 min. Eighty percent of the deforestation is concentrated in four provinces: Santiago del Estero, Salta, Formosa, and Chaco [23].

At the same time, variations in local and regional climate had begun to be noticed in the country. The average annual temperature increased from 1960 to 2010 in almost all the northwest subregions (and Cuyo); in many areas (more than 0.5°C), the most notable changes were observed in spring. From 1950 to 2010, the annual average temperature, through the region was 0.6°C and it reached 0.7°C in Salta and Jujuy [22]. At a national level, the average temperature increases from ½ to 1°C. The possibility of increasingly intense heat waves has been forecast. In the northwest, an increase of 4–5° is projected by 2030, one of the highest on the planet. In the west and, notably, in the north of the country, there has been a shift toward the extension of dry winters. This could be generating problems with water availability for the populace, more favorable conditions for wildfires in forests and grasslands, as well as stress on livestock. This could bring implications on the biodiversity of the native forest remnants in

Yungas [22], and, at the same time, the disappearance of such remnants, which could provide feedback for those changes that are taking place at an atmospheric level.

Improving the understanding of biomass and carbon stocks in forests, therefore, provides valuable information for use land planning and designing comprehensive strategies in the context of global climate change. The purpose of this chapter is to present a synthesis of some of the different works developed in the subtropical forest of the Pedemontana Jungle, based on years of studies in the area. Studies were focused on the northern of the country, noting its role as carbon reservoirs and discussing factors that could influence the carbon sequestration total capacity of the same. The information presented here, without doubt, will contribute to the construction of a reliable database of this potential, which will facilitate standardization of units, reduction of uncertainties, and contribution to a more efficient strategy to limit GHG emissions, providing some learning and useful recommendations. Inasmuch as this ecosystem extends to Venezuela, the results obtained will provide a frame of reference for future studies on this ecological zone. This information is also necessary to improve the understanding of the distribution patterns of biomass and carbon at the global level and to describe patterns of land use. The results presented could guide in designing plans and management policies for these types of forests, at national and international levels.

2. Materials and methods

2.1. The Yungas ecosystem: Pedemontana Jungle

The phytogeographic Yungas province borders the Andes mountain range from Venezuela to Argentina [24]. The Argentine Yungas, which constitute a vital habitat for the fundamental role in the regulation of the water basins and protection against erosion, have been subjected to a long history of anthropogenic interventions, especially in low-lying areas, called the Pedemontana Jungles, which have a high agricultural potential [25].

The history of Pedemontana Jungle in the north of Argentina has been closely tied to the railway expansion, necessary for the transport of precious wood, tropical crops, and sugar. More recently, from the 1990s, soybeans won the major role, expanding rapidly in the foothills landscape and its transition to the Chaco plain. The deterioration from the advance of the agricultural frontier, coupled with logging, the commercial bird catching and poaching – among others – are causes for concern because of the almost 5 million hectares that cover the Argentine Yungas, the effectively protected area is only 5% of the total [21].

The Pedemontana Jungle, which stretches from 450 to 900 m.a.s.l. – other authors mention minor ranges [24] – and which represents 25% of the Yungas, has been considered as an ecosystem in danger of extinction, and its deforestation would eliminate 30% of the total Yungas biodiversity [25]. In this region, 120 species of mammals and 8 of the 10 species of neotropical cats are represented. Also, approximately 583 species of birds inhabit it, which represent 60% of the species in Argentina [26]. Likewise, in the Pedemontana Jungle of Argentina and Bolivia, they were identified 18 AICBA (Areas of Importance for the Conservation of the Birds of Argentina), noting that the AICBA including sectors of the Pedemontana Jungle, have a diversity of birds comparable to the cloud mountain forests (ecological zone of higher altitude than the PJ) and higher than the Chaco forests that surround them [27].

The physiography varies from submountain foothills to alluvial descents, presenting a hilly and wavy topography. The soils present in the study area are according to the taxonomic classification of FAO soils of the Phaeozem Haplic and Luvic type [28]. Soils of luvisol calcium were recorded only on the Coronel Moldes site.

The climograph and altitude for each of the studied sites are shown in Figure 1.

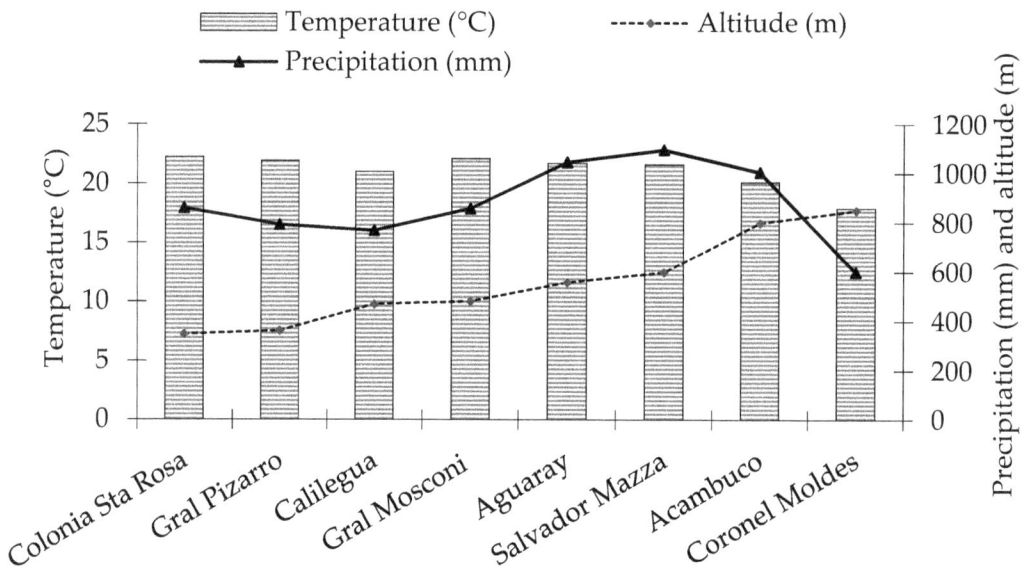

Figure 1. Annual average rainfall (mm), annual average temperature (°C), and altitude (m.a.s.l.) for different sites studied in the Pedemontana Jungle in northern Argentina (Salta and Jujuy provinces). Source: http://es.climate-data.org/location/145171/

2.2. Case studies

All the studies summarized in this chapter were carried out in the province of Salta, in northern Argentina, with the exception of case II, which was developed in the province of Jujuy. The province of Salta has an area of 155,500 km², occupying the sixth position at the national level, and with a value similar to the surface of Nepal, has a population of 1.2 million, making it the eighth most populous out of 23 at the national level.

Of the entire surface occupied by the Yungas ecosystem in the country, 61% of it extends through this province, making it essential to focus on studies in this particular region. Also, Salta has 23% of the total surface of the country's native forests, and deforestation in this province is triple the world average [29].

Some of the assumptions, which have been evaluated from various case studies (always focusing on the Pedemontana Jungle), are as follows (more detailed in Table 1):

i.   The subtropical rainforests of the country have a greater capacity for carbon sequestration than subtropical dry forests at identical latitude.

ii.   Carbon sequestration in forests disturbed by human activity is lower than in forests less seized by humans, releasing the difference of carbon into the atmosphere.

iii. The carbon stock, in legally protected forest sectors, is higher than in other sectors without protection located at identical latitude and under similar conditions.

iv.   The potential for carbon sequestration in the Pedemontana Jungle is less if latitude increases.

v.   The fragmentation of the Pedemontana Jungle generates microclimatic changes at the edges, which could affect carbon sequestration.

Case	I	II	III	IV	V
Legal protection	No	Yes	Yes and no	No	No
Site	Coronel Moldes	National Park Calilegua	Wildlife Reserve of Acambuco and Campo Pizarro	Aguaray and General Pizarro	Colonia Santa Rosa
Plot number	23 main plots (AGB_{10}) for each ecosystem; 23 plots of 50 m^2 (AGB_0); 23 plots of 5 m^2 (LUV); 46 plots of 1 m^2 (HUV and LI); 138 soil plots (SOC)	20 main plots (AGB_{10}); 20 plots of 50 m^2 (AGB_0); 40 plots of 1 m^2 (HUV and LI); 120 soil plots (SOC)	50 main plots (AGB_{10}); 50 plots of 50 m^2 (AGB_0); 250 soil plots (SOC)	50 main plots (AGB_{10}); 50 plots of 50 m^2 ($AGB0_0$); 250 soil plots (SOC) and 500 microclimatic instantaneous records (MIC)	78 main plots (AGB_{10}); 78 plots of 50 m^2; 468 soil plots (SOC); 156 microclimatic instantaneous records (MIC)
Carbon pool	AGB_0 AGB_{10} LI HUV LUV BGB SOC	AGB_0 AGB_{10} LI HUV BGB SOC	AGB_0 AGB_{10} BGB SOC	AGB_0 AGB_{10} BGB SOC MIC	AGB_0 AGB_{10} BGB SOC MIC

The acronyms AGB_{10}, AGB_0, HUV, LUV, LI, SOC, MIC are explained in the text.

Table 1. Methodological differences and similarities between the case studies.

2.3. Sampling design

The methodology used for each of the case studies presented in the next section shows some differences that are summarized in Table 1. In most cases, the data were collected following a random sampling design. Only in case V, the sampling was systematic.

The experimental design used was nested plots. Main plots had a total area of 100 m^2 and were rectangular plots. The criterion used to determine sample size for each stratum was an estimation of AGB of trees with a diameter at breast height (dbh) $\geq$ 10 cm during pre-sampling (90% probability and 20% mean standard error).

Carbon represents about 50% of the total oven-dried biomass present in forests [32]. Estimation of carbon pools in forests necessarily involves studying the different strata of biomass present in them. In the different studies, the following carbon pools and variables were measured:

a. Aboveground tree biomass: AGB refers to the total amount of aboveground living organic matter in trees and shrubs ($\leq$1 cm dbh and $\geq$50 cm height) expressed as oven-dried tons. Total height (from ground level up to crown point) and dbh were measured in all trees with dbh $\geq$ 10 cm (called AGB$_{10}$) in 100 m^2 plots. When $1 \leq dbh \leq 10$ cm and height ≥ 50 cm (called AGB$_0$), trees were measured in 50 m^2 plots. In multiple-stemmed trees, only the longest stem was measured. If neither shoot was dominant, an average of similar shoots was calculated. The basal diameter was registered only when the stem was shorter than the dbh. Standing dead trees with dbh $\geq$ 1 cm and fallen trees with dbh $\geq$ 10 cm were measured in the same way as living trees. However, correction factors of 0.8 and 0.7, respectively, were applied to the biomass values obtained. For hollow or ill trees, a factor of 0.9 was applied.

b. Lignified understory vegetation (LUV): All shrubs shorter than 50 cm were collected in 5 m^2 plots within the corners of the main 100 m^2 plots.

c. Herbaceous understory vegetation (HUV): This fraction was removed in two 1 m^2 plots. These plots were located in opposite corners within the 100 m^2 plots used to measure AGB$_{10}$.

d. Litter (LI): Organic debris on the soil surface (including freshly fallen parts of plants, decomposing organic matter, and deadwood) with a diameter no greater than 10 cm were collected in the same plots used for HUV.

e. Belowground biomass (BGB) (tree roots): Due to the difficulties involved in the measurement of this fraction, it was estimated indirectly as a proportion of AGB$_{10}$ for Chaco and Yungas.

f. Soil: Bulk density and percentage of organic carbon were determined in soil samples collected at a depth of 30 cm [30]. Vegetation and litter were removed from the soil surface prior to sampling. Bulk density was determined in two samples per plot using the cylinder method. Results from these samples were averaged. The percentage of organic carbon was measured following the method described in Walkley and Black. This measurement was performed on a composite sample built from four samples taken at identical distances within a linear transect along the longest axis of the 100 m^2 plots (dimensions of these plots were 5 m × 20 m).

Wet weight was recorded on site for LUV, HUV, and LI fractions. Dry weight was determined in the lab (registered after drying in an oven at 80°C until constant weight). The equation introduced by Cairns and coworkers [31], for tropical forest and lower latitudes than 25°, was used. The AGB fraction, also called as "biomass density" when expressed as tons of oven-dried weight per ha [32], is the main source of total biomass in a forest ecosystem. Its relevance as a GHG mitigation option is therefore crucial [11–13]. This fraction was thoroughly assessed using a nondestructive methodology: allometric equations (Table 2).

g. Solar radiation intensity (W/m^2): A LICOR 250 pyranometer was used with a silicon sensor with a resolution of 0.1 W/m^2. The measures of global radiation readings are precise to ±5%.

h. Air relative humidity (%): This was recorded using a psychrometer or hygrometric probe Vaisala HM 34. Reading is immediate and accuracy is ±2%. The sensor is the Humicap type. Measurements were taken at 1.5 m from ground level.

i. Air relative temperature (°C): This was registered with a Vaisala HM 34 probe, with temperatures ranging from −20 to +60°C. Measurements were taken at 1.5 m from ground level.

j. Soil temperature (°C): This was measured with a FLUKE 54 II digital thermometer with accuracy ranging from 0.05% + 0.3°C. Measurements were taken at 10 cm depth.

k. Soil humidity (%): This was estimated by two soil samples taken at 10 cm depth per plot.

Authors	Carbon pool	Equation	No.
Chave et al. (2005)	AGB_{10} and AGB_0	$AGB = \exp(-2.977 + \ln(S.D^2.H))$	(1)
Brown et al. (1989)	AGB_{10} and AGB_0	$AGB = \exp(-2.4090 + 0.9522 \ln(S.D^2.H.))$	(2)
Chave et al. (2005)	AGB_{10} and AGB_0	$AGB = 0.112(S.D^2.H.)^{0.916}$	(3)
Gehring et al. (2004)	AGB	Diameter (30 cm) $= 1.235 \times dap + 0.002 \times (dap)^2$ $AGB = \exp(-7.114 + 2.276*\ln(D_{30}))$	(4)
Cairns et al. (1997)	BGB	$BGB = \exp(-1.0587 + 0.8836 \times \ln(AGB))$	(5)
Macdicken, (1997)	SOC	$SOC = OC \times BD \times D$	(6)
Sevola (1975)	V	$V = -2.2910 + 0.0558 \times \log (D^2 \times H)$	(7)
Sevola (1975)	V	$V = -3.2794 - 0.0734 \times \log H^2 + 1.0580 \times \log(D^2 \times H)$	(8)
Sevola (1975)	V	$V = -2.4385 + 0.9560 \times \log(D^2 \times H) - 0.80350 \times \log(H / D)$	(9)
Chave et al. (2014)	AGB_{10} and AGB_0	$AGB = 0.0673 + (S.D^2.H)^{0.976}$	(10)

AGB = tree aboveground biomass (kg oven-dry); SB = stem biomass; S = wood density (oven-dried biomass per green volume, in t/m³); D = diameter at breast height (1.3 m above ground, in cm); D30 = diameter at 30 cm above ground; H = total height (m); BGB = belowground biomass (t/ha); OC = concentration of organic carbon in the soil (%); BD = soil bulk density (g/cm³) and D = depth of soil (cm); V = total tree volume, in dm³, included stem, bark, and branches.

Table 2. Allometric equations used in this chapter.

The last five parameters called "microclimate factors" were measured in each preset distance, for each transect study, always at midday between 12 p.m. and 2 p.m. In the case of values per site, the different measurements taken were averaged per plot.

2.4. Estimation of biomass and carbon

Once field measurements were carried out, the data were computed clerically, carrying out the biomass estimate for each compartment, transforming it into carbon values (factor of 0.5 [32]) and achieving the sum of all the carbon pools. All equations used are shown in Table 2. Equation (1) was developed by Chave et al. [33] for "moist forest stand," while equation (3), by the same authors, was developed for "dry forest stands" (applied to the Chaco). Equation (10) was recently developed by these authors and was applied to the *Anadenatnhera colubrina* and *Cedrela angustifolia* species, for which no specific equations were found. Equation (4) was applied only in vines and required converting the dbh into diameters at 30 cm height, and then entering that value into the equation [34]. In the case of volumetric equations (7, 8, and 9) [35], the total biomass conversion was carried out by multiplying the total volume by the basic density of each species. Equation (7) was then applied to the *Calycophyllum multiflorum* species, equation (8) to *Phyllostylon rhamnoides*, and equation (9) to *Astronium urundeuva*, all equations being developed in the region.

The basic wood densities (dry) for different species were obtained from Ref. [36]. A basic density value obtained from the weighted average of the densities of each site's species was used for the species that for various reasons could not be identified. For estimation of SOC (soil organic carbon), equation (6) was used [30]. For data analysis, the nonparametric type test was chosen. We used the INFOSTAT® software, and a value of 0.05 was considered significant.

3. Results and discussion

3.1. Effect of temperature and humidity on the carbon stock: dry and humid subtropical forests at the same latitude

Contributors: Manrique, S.M. and Franco, J.

The subtropical moist forests of the country have a greater capacity of carbon sequestration than subtropical dry forests at identical latitude.

As was mentioned, the Chaco ecosystem is the largest surface area at the national level. It was interesting to compare facets of this ecosystem with the Yungas Pedemontana Jungle with regard to the potential for carbon sequestration at the same latitude. The work was carried out in the municipality of Coronel Moldes (25°16′00″ South latitude and 65°29′00″ West longitude), 60 km south of the capital of the province of Salta.

The province's climate is defined as subtropical mountainous with a dry season. However, the topography does allow the development of contrasting environments. Thus, the moist winds from the southeast enter the province and release their moisture from submountainous ranges that make up the sub-Andean hills in the north-central region of the country. This allows the spread of vegetation, which is a unique environment that runs along elevations in different altitudes, forming a north–south strip. The Chaco ecosystem develops on the plain that extends from the center of the country to the East, and in Salta two districts are exhibited: the semiarid Chaco and the mountain Chaco. Precipitation decreases as it moves eastward, shrinking from more than 650–700 mm per year in the Pedemontana Jungle ecosystem to values of less than 460 mm in the Chaco ecosystem. Temperatures also suffer a slight increase as it moves west away from the mountains, which have the moisture [37], marking isotherms in the range of tenths of degrees, as the distance between the mountains and the eastern point increases.

The starting points are corroborated in this study: the most humid ecosystem shows a carbon stock 43% larger than that stored in the driest ecosystem (Table 3). In the case of Yungas, the AGB fraction means almost 80% of the total biomass, although the greater fraction (AGB_{10}) alone implies 71%, leaving the AGB_0 a reduced participation. The BGB means more than 16% of total biomass, and the rest if divided between the LI (about 3%), the LUV (with almost 2%) and lastly, the negligible participation of HUV (0.1%). For the Chaco, the fraction AGB provides more than 71% of the total biomass, where the trees of larger diameters (AGB_{10}) mean 66% of this contribution. In this environment, the BGB takes on greater importance (with more than 21%), and is followed by – in the identical order shown in the Yungas environment – LI fraction (3.2%), LUV (2.6%), and HUV (0.6%) (Figure 2).

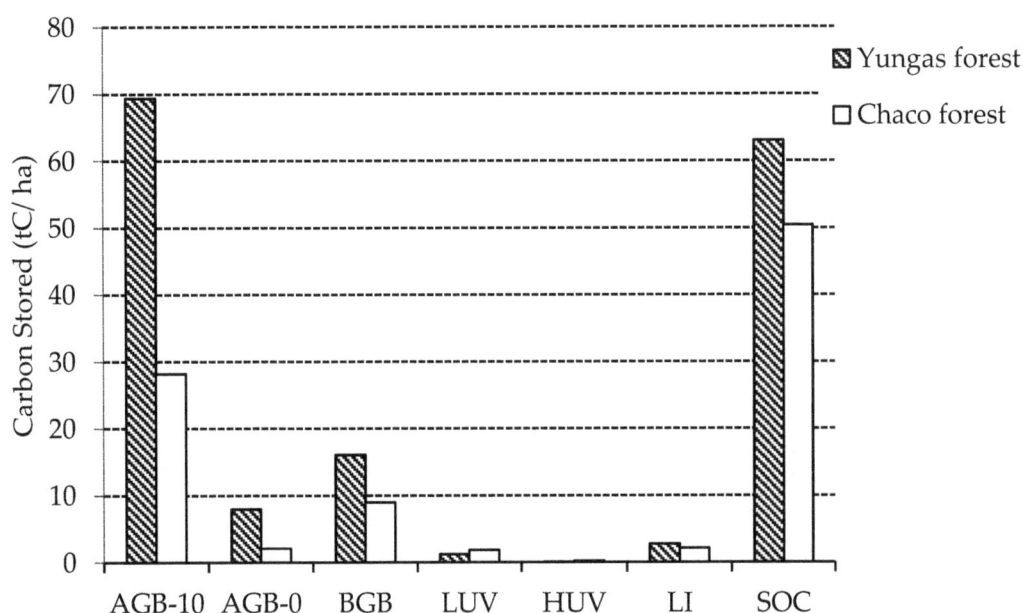

Figure 2. Carbon stock and contribution of each carbon pool studied. The acronyms AGB_{10}, AGB_0, BGB, HUV, LUV, LI and SOC are explained in the text.

Clearly, the AGB and SOC fractions are the two largest contributors in the two ecosystems. In Yungas, the AGB represents 48% of the total fixed carbon, while the SOC contributes 39%. In the case of the Chaco, 33% of the total carbon in the ecosystem is concentrated in AGB, while

54% remains captured in SOC. Soil is an important reservoir of carbon, becoming the most important fraction in dry environments. However, when we compare the absolute values of SOC in both environments, the soil shows a significant relationship with the vegetation found on the surface. In Yungas, it is 63 tC/ha, while in Chaco it is 50 tC/ha.

Sector	Average	Standard deviation
Yungas forest (Selva Pedemontana)	162	85
Chaco forest (Chaco Serrano)	92	42

Table 3. Carbon stock (tC/ha) in both ecosystems, Chaco and Yungas, in Coronel Moldes, Salta, Argentina.

Viglizzo and Jobbágy [21] point out that the carbon stocks in the biomass and in the organic fraction of the soil in Argentina vary from one ecoregion to another. The carbon stock in biomass is directly associated with the availability of vegetation biomass. In the tropical and subtropical regions of Argentina (e.g., Yungas), more than 50% of total carbon is found stored in the AGB fraction, which makes this element vulnerable and easily appropriable by humans. This relationship falls dramatically in areas dominated by grasslands/pastures (e.g., Chaco), and even more (without reaching 10%) in intensively cultivated ecosystems.

In Yungas, the average height was 11 m and average dbh was 17.6 cm, both higher than those for Chaco, although still lower than figures cited for pristine Yungas ecosystem [24, 25, 38]. Estimations made for tropical humid forests around the world range from 150 to 192 t/ha for closed, undisturbed forests and around 50 t/ha for open forests [39]. Certainly, different factors may be influencing these differences (rainfall, soil type and site features, topography, etc.) [32, 33, 39, 40]. Moreover, the structure of the forest in the Yungas area included in this study was clearly disturbed by humans and livestock. Numerous recent and decomposing stumps were found and there were unambiguous signs of wandering animals and persons. *Solanum riparium* was also abundant in this area, a species normally dispersed by wild animals or cattle. The appearance of typically Chaco species in sections of Yungas forest is probably a sign of human intervention in this region [24, 38].

Our results suggest that forest degradation is detectable not only in Yungas but also in Chaco. In environments similar to Chaco, discrepancies between these results (lower) and estimations made in similar environments in other forests of the world might be due to structural differences, altitude, latitude and humidity, gradients (24, 32, 33). However, in our case the level of degradation exerted by human activity in this environment might also be responsible for the discrepancies [20, 21, 41] (further details refer to [43]).

Economic activities such as agriculture and logging, which take place in these ecosystems, are arguably not respecting their carrying capacity. Local institutions do not seem to be capable of stopping, controlling, or regulating these activities. Whether entering into a market-based system like the one promoted by the Kyoto Protocol will be part of the solution to the problem of deforestation and conservation of local native forests remains to be seen. Decisions are highly political and many times the relevant decision makers are thousands of kilometers away. No decisions affecting the future of these forests should be taken until agreements on this issue are reached or until judiciary processes are properly finished. Competing claims on the ownership of the forestland, the products of the forests, and the provision of ecosystem services must be taken into consideration in a comprehensive forest management.

3.2. Effect of human influence on the carbon stock in forests

Contributors: Gallucci, G.B. and Manrique, S.M.

Carbon sequestration in forest disturbed by human activity is lower than in forests less seized by humans.

In case I, we identified that studied forest sectors clearly show human influence as a factor of degradation of the original structure of the same type. In this case study, it was interesting, particularly, to assess this difference and try to quantify it for samples of the same Pedemontana Jungle ecosystem, but this time as a protected area: Calilegua National Park (23°27'–23°45' South latitude and 64°33'–64° 52'0" West longitude). The park was created in 1979 to protect a representative sector of the Yungas and to protect the headwaters of the Calilegua streams, which are a part of the San Francisco River basin, and provide water to neighboring crops in the protected area. With an area of 76,320 ha, it is the largest national park in the Argentine Northwest. It is approximately 165 km from the city of Salta.

We studied two areas of the park (north and south sectors) separated by only 50 km but which have different accessibility to human influence. The north sector surrounding the town of Caimancito has been invaded by oil companies, which have conducted exploration activities in the area, and therefore have dissected the forest, leaving open "choppings" or paths of prospecting. This has led to the accessibility of nearby residents who have taken advantage of the forest and even have led their animals to graze there. In the south, on the other hand, exploration activities were not carried out and therefore, even if villagers could have accessed the site, on its more sheltered side (the other side of rivers that flow through the park), a better conservation has been maintained, which can be seen in the large, heavily wooded trees, and the high forest value that is still there. Surely, the presence of Park Rangers (Aguas Blancas section) in this sector has helped much in this protection.

Two sectors that maintain homogeneous topographic, edaphic, and climatic conditions were selected. Both sectors were compared through analysis of average annual rainfall records (56 years series) without finding statistically significant differences ($H = 0.01$, $p > 0.999$). Records of minima and maxima were also analyzed. The series of annual average temperatures were not statistically different ($H = 0.16$, $p = 0.686$). In the case of edaphic variables, existing cartographic studies allowed us to associate both sectors with the same series of soils. Organic matter samples taken in the area showed no significant differences ($H = 4.71$, $p=0.210$). It was assumed that both sectors had identical site conditions.

We evaluated the same carbon pools as in case I with the exception of LUV, which had no relevant participation in the previous case, and therefore it was not included in the pursuit of reducing the fieldwork effort and costs.

The obtained results allow us to advance with the basic assumption: the north sector, subject to anthropogenic influence, it showed a carbon stock 23% lower than the south sector, which had less accessibility and a better state of conservation (Table 4). These differences were statistically significant ($H = 11.20$, $p < 0.001$) only for the AGB stratum, but not for the other strata studied nor for the total carbon stock. Under similar conditions of climate, soil, geomorphology, altitude, and latitude, the human influence could explain these differences, as the AGB stratum is the easiest to appropriate by humans [10, 17, 19, 21]. The AGB make the largest contribution in both sectors to the carbon stock (53, 55%), followed by SOC (28–31%) and finally BGB (8–10%) depending on the sector analyzed (Figure 3).

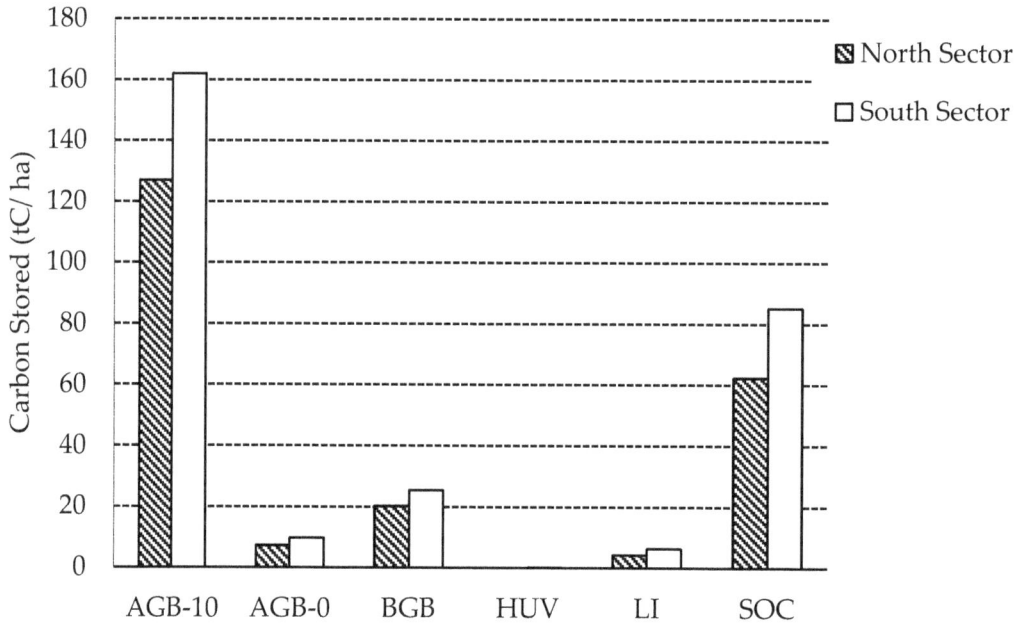

Figure 3. Carbon stock and contribution of each carbon pool studied. The acronyms AGB_{10}, AGB_0, BGB, HUV, LI and SOC are explained in the text.

Sector	Average	Standard deviation
Protected forest degraded (north sector)	221	116
Protected forest better preserved (south sector)	272	129

Table 4. Carbon stock (tC/ha) in both sectors, north and south, in Calilegua, Jujuy, Argentina.

Against the results, there is an urgent need to review the administration and safeguards for the Calilegua National Park, with a reinforcement of the Corps of Rangers in the area (currently with few people that must patrol the whole park). Other authors are agreed that the declaration to protect does not always mean adequate protection [43, 44]. The acquisition of more financial resources for the protected areas should be carried out in the light of a strict management plan and monitoring. Poaching, livestock grazing, and logging without authorization – with the thinning out of valuable wood species – must be eradicated from the core area, so that the Park can fulfill its role with the conservation of biodiversity, which has been included in the international statement "Yungas Biosphere Reserve."

3.3. Effect of legal protection on the ecosystem

Contributors: Manrique, S.M.; Vacaflor, P.; Fernández, M. and Franco, J.

The carbon stock in legally protected forest sectors is higher than in unprotected sectors located at the same latitude and under the same conditions.

In case II, two sectors of the legally protected Pedemontana Jungle were analyzed, which clearly show differences between them in their accessibility to human influence. It became interesting to continue in this line of study, exploring if the trend found in the former case could be due to a particular situation in the Calilegua National Park. In this case study, we sought to observe comparative sectors inside and outside legally protected regions located at the same latitude and altitude, and under the same conditions. We started to identify protected areas in the province which shelter samples from the Pedemontana Jungle. We finally worked in and out of the Provincial Reserve of Flora and Fauna of Acambuco (PRFFA) (22°12'38.5" South latitude and 63°56'23.1" West longitude) and in the National Reserve of Campo Pizarro (NRCP) (24°11'54.87 and 24°14'21.7" South latitude, and 64° 7'27.00" and 64° 9'23.79" West longitude). The creation of PRFFA dates back to 1979, and currently has an area of 32,000 ha. It is approximately 470 km from the city of Salta to PRFFA. In the case of NRCP, it was created in late 1995 with an area of 25,000 ha, and soon after a process of reversal and social conflict, the NRCP ended up with an area of 21,000 ha. It is approximately 280 km from Salta. In this study, efforts were concentrated in the carbon pools considered most significant in the prior cases, eliminating HUV and LI from the samples.

The results show that, on average, the carbon stock is similar in protected and nonprotected areas (Table 5). Having considered the average of all sectors included in the Reserves and the average of all the studied sectors not protected in them, no significant differences were found ($H = 0.85$, $p = 0.356$), by even analyzing just AGB separately ($H = 0.98$, $p = 0.322$). The initial assumption cannot be confirmed: no case shows that the legal protection has caused differences in the ecosystem it protects, neither favoring nor against. Yet we see different values if we consider the samples of the north sector and south sector separately, as will be discussed in the following section.

Sector	Average	Standard deviation
Protected forest	203	74
Unprotected forest	213	82

Table 5. Carbon stock (tC/ha) in both sectors, protected and unprotected forest, in Acambuco and Campo Pizarro, Salta, Argentina.

In terms of the importance of each of the studied carbon pools (Figure 4), the carbon fixed at the fraction of AGB returns to be larger than the fixed carbon in the soil (SOC).

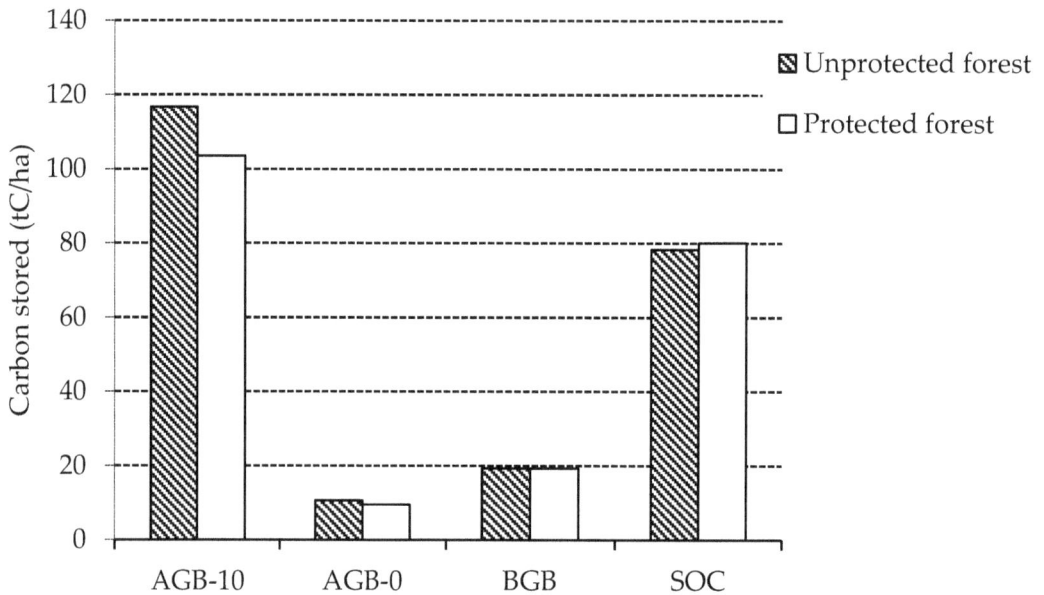

Figure 4. Carbon stock and contribution of each carbon pool studied. The acronyms AGB_{10}, AGB_0, BGB and SOC are explained in the text.

The inclusion of Pedemontana Jungle sectors within legal protection figures has not resulted in benefits in terms of their ability to sequester carbon in the different carbon pools. However, this consideration is not conclusive in the role of protected areas. Pedemontana Jungle sectors, within and without the protected areas, could be similar in their capacity to sequester carbon in two possible situations: (i) a good general ecosystem condition level, which still remains a certain continuity of ecosystem, and therefore, either inside or outside the Reserve, is of similar forest samples and show no particular features nor different structural configurations; (ii) a poor state of conservation, which has equally affected protected and nonprotected areas, imprinting similar features in the different sectors, by simultaneous intervention in the different forest sectors. A more in-depth study of other ecosystem variables would perhaps lean toward one alternative or another. However, the different log of measured carbon stock in case II in Calilegua National Park, or the degradation features in the two types of ecosystems (Yungas and Chaco), which were observed in case I, clearly indicates that the forests have not received the attention they should have over the years.

Therefore, these thoughts should be a trigger to continue with deeper and more comprehensive evaluation and to draw attention to the need to review and update control schemes and monitoring of native forests – mainly protected areas. The global community has recognized the importance of forests for biodiversity, and has prioritized the preservation of forest biodiversity and ecosystem functions through multiple multilateral agreements and processes. For example, the Aichi Biodiversity Targets established by the Convention on Biological

Diversity (CBD) in its strategic plan include halving the rate of loss of natural habitats including forests (target 5) and conserving 17% of terrestrial areas through effectively and equitably managed, ecologically representative, and well-connected systems of protected areas (target 11). Currently, designating protected areas is one of the primary strategies for conserving biodiversity. Different authors have discussed the increase in protected areas over the past century; however, they find that many key biodiversity areas are not adequately covered by protected area status [44].

The always-protected system areas will be limited to preserve all the original diversity, but even so, it is imperative that these areas exist and continue to expand with scientific criteria.

3.4. Effect of latitude and altitude on the carbon stock

Contributors: Manrique, S.M.; Vacaflor, P. and Fernández, M.

The potential of carbon sequestration in the Pedemontana Jungle is less if the latitude increases.

In case III, the average carbon stock in protected and unprotected areas was approximately similar, although with different values from cases I to II. This led to the analysis of the position within the ecosystem of the studied sites. In case I, with 162 tC/ha, the forest is at 25°16 and 65°29. In case II, with 221–272 tC/ha, the forest is approximately between 23°27 and 64°33. Analyzing other sectors located at different latitudes could confirm the trend of higher values of biomass and the north sector of the Pedemontana Jungle (e.g., Calilegua showing a value 100% greater than the Coronel Moldes value) and values that decrease toward the south. It was interesting, therefore, to explore case III results separately, taking as a northern sector, the plots carried out near Aguaray (22°12 and 63°56) and, as a southern sector, those carried out near General Pizarro (24°11 and 64°7).

Sector	RI (W/m²)	RH (%)	RT (°C)	SM (%)	ST (°C)
North sector	0.065 a	43.55 a	26.94 a	9.9 a	22.09 a
	(0.02–0.44)	(21.7–86.7)	(16.6–36.55)	(3.15–17.25)	(20.15–25)
South sector	0.065 a	36.9 a	31.66 b	7.24 ab	24.2 b
	(0.02–0.13)	(16.35–55)	(25.4–37.8)	(2.3–14.4)	(22.85–30.8)

RI = radiation intensity; RH = relative humidity; RT = relative temperature; SH = soil moisture; ST = soil temperature. Mean and range for each variable. Means followed by different letters (a, b) within the same column indicate statistically significant differences (P <0.05).

Table 6. Average climatic conditions.

The loss of species diversity and conditions of humidity and altitude from north to south, along the gradient in which the Pedemontana Jungle extends within Argentina, has been previously documented [27, 40]. Therefore, it was interesting to know in this case if this trend was also clearly reflected in the carbon stocks of the studied sectors of the Pedemontana Jungle. If the previous studies, the participation of the HUV and LUV strata was between 0.01% and 0.02% and the LI carbon pool was between 1.5% and 3%. Therefore, in this study efforts were concentrated in the strata of AGB, BGB, and SOC. The chosen sectors show the average weather conditions (for the same season, day, and year), which differ significantly in air relative temperature (RT), moisture and soil temperature (SM and ST, respectively) (see Table 6).

Studies of carbon stock results show that the two sectors are clearly separated in terms of their potential. The northern sector has the largest records of total carbon with an average of 242 tC/ha, while the southern sector registers an average 28% lower (Table 7) with statistically significant differences (H = 12.38; p < 0.01). These values can be associated with different microclimates, possibly generated by a latitude effect, whose influence on climatic variables can be seen in Table 6. In all cases, the differences are in favor of a cooler, more humid climate in the northern sector and warmer and drier in the southern. Although the number of analyzed sectors in a latitudinal gradient in the Pedemontana Jungle (narrow strip of north–south direction), are not representative of the whole distribution, the data can be interpreted in light of the existing scientific studies in the area [25, 27, 40].

Once again, the two carbon pools that make a greater contribution to the total ecosystem carbon stock are AGB and SOC, being greater in the case of the northern sector, meaning 52% and 34% of the total carbon stock, respectively (Figure 5). This implies that more than 86% of total carbon is concentrated in these two fractions. In the southern sector, the participation of these carbon pools is 47% and 38% for AGB and SOC, respectively, but with greater involvement of the SOC carbon pool in this case.

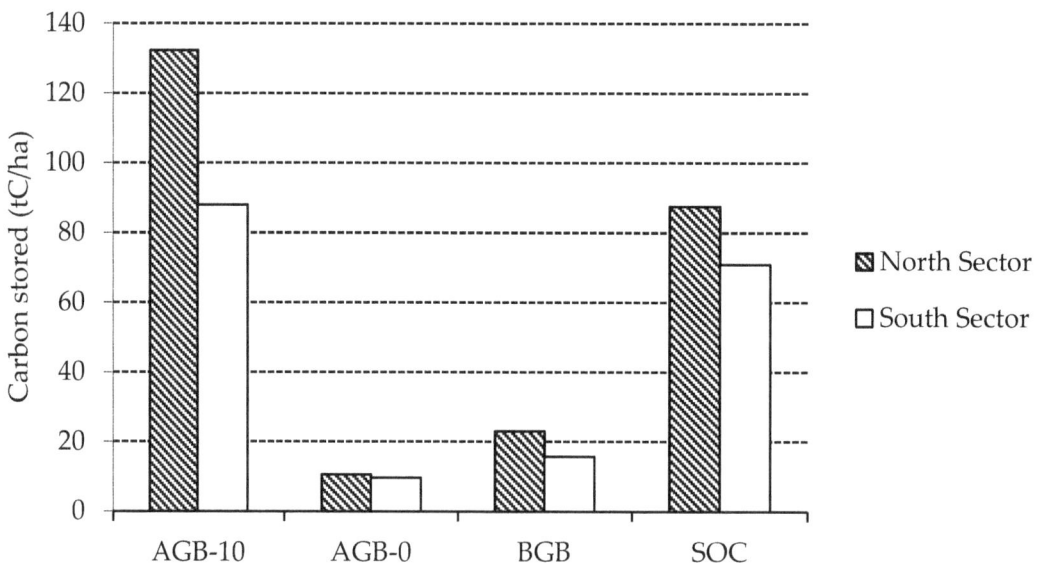

Figure 5. Carbon stock and contribution of each carbon pool studied. The acronyms AGB_{10}, AGB_0, BGB, and SOC are explained in the text.

Sector	Average	Standard deviation
Forest in the north (22° latitude)	242	96
Forest in the south (24° latitude)	174	68

Table 7. Carbon stock (tC/ha) in both sectors, north and south, in Salta, Argentina.

However, always considering the altitude of the Pedemontana Jungle, as the latitude increases, the altitude decreases in general terms. It has been recognized that the floral changes are influenced by complex interactions of weather and edaphic variables in Yungas altitude ranges [24, 25, 40]. Beyond the fact that the associated variables of increasing altitude (which in this study varies between 22° S and 24° S) and/or altitude (which varies between 500 and 700 m.a.s.l. in the southern sector and between 700 and 900 m.a.s.l. in the north) would more or less determine overt changes at the level of species and ecosystems, such variations exist without doubt, and they are defining two sectors of the same forest in terms of carbon sequestration potential.

These observations indicate that it is essential to preserve sectors of different latitudes and altitudes in the Pedemontana Jungle, since there are intrinsic factors that are defining differential features in the biomass and carbon stock, as well as, in every ecosystem functions associated with these particular conditions [40]. Other authors have already pointed out that the recommendation in all cases is to maintain connectivity of Yungas in distribution, safeguarding different sectors of the Pedemontana Jungle, varying in latitude and altitude [27].

Human influence, not analyzed in this study, will, no doubt, imprint differential features over time if their presence is not restricted, since we have observed signs of livestock and logging in the different studied areas. In the southern sector, where the Pedemontana Jungle has been deeply fragmented and immersed in an array of crops, it is considered that there might be a microclimatic influence on the fragments by the existence of rough edges [7, 18]. This aspect will be dealt with in the following section.

3.5. Effect of fragmentation on the carbon stock

Contributors: Manrique, S.M.

Fragmentation of the Pedemontana Jungle generates microclimate changes at its edge, which could affect the sequestration of the carbon stock.

The fragmentation of forests, reducing surface and insulation, exposes organisms, which remain in the fragment, to conditions differing from their ecosystem, which is primarily manifested in the contact between two different environments, which has been defined as "edge effect" [18], and that impact toward the forest interior.

Microclimatic changes caused as a consequence of contrasting conditions between the remnant forest and the adjacent field, subjected to different uses (cultivation, planting, and pastures), would seem to be the most immediate and apparent fragmentation changes [7]. Several authors have recognized that, at the edge of the fragments is an environmental gradient toward the interior: generally brightness, evapotranspiration, temperature, and wind speed decrease, while soil moisture and humidity increase toward the interior of the fragment. Biological changes could then arise as a result of these changes in the microclimate of the fragment edges [7, 18].

This study sought to analyze and quantify the possible microclimatic changes generated in the fragment edges of the Pedemontana Jungle, also observing the distribution of five representative tree species (by their frequency [24]). The studied species were as follows: (i) *Calycophyllum multiflorum Griseb, Castelo*, (ii) *Phyllostylon rhamnoides J.Poiss., Taub*, (iii)*Astronium urundeuva Engl.*, (iv) *Anadenanthera colubrina Vell., Brenan,* and (v) *Cedrela angustifolia DC.* It was estimated that the typical species, "climax" or more conservative ones of the population (e.g., those that have higher demands for their germination or growth requirements and with low tolerance for humidity fluctuations), could be more easily eliminated like those selected for this study. These species, which have a high degree of integration, complexity, and efficient energy use, are recognized as more susceptible to edge changes [18]. Therefore, in fragmented environments, the survival advantage is given to those pioneer species with a maximum tolerance for a wide range of environmental conditions.

Five forest sectors in the Colonia Santa Rosa municipality were worked (23°20'00 south latitude and 64° 30'15" west longitude): four, clearly turned into fragments, and one continuous (not fragmented) taken as a standard for comparison. The fragments were of distinct sizes: two large (sites 1 and 2 between 160 and 180 ha) and two small (3 and 4 between 3 and 5 ha). The distance from the city of Salta is 250 km.

The results of microclimatic records (taken from the edge toward the inside of the fragments, except in the site 5 as it was not considered the same edge but worked in an inside sector, looking for original ecosystem conditions) suggest that (Figure 6):

- High radiation intensity (RI) values are recorded at the edge (around 800 W/cm^2 on average) and almost constant values under cover (forest interior), which mean, almost, only up to 2% of that value. The differences were statistically significant ($H = 16.19$; $p < 0.01$).

- Soil moisture (SM) in the interior is twice that of the edge (with maximum values of up to 16%) showing significant differences ($H = 29.20$; $p < 0.001$).

- Air relative humidity (RH) increases toward the interior, reaching values up to 7 times higher than those at the edge (up to 53% relative humidity (RH)). The differences are significant ($H = 5.41$; $p = 0.048$).

- Soil temperature (ST) is one of the most stable variables, although differences can be detected: considering 100% at the edge (18°C on average) is reduced to 25% in the interior. The differences are not statistically significant ($H= 5.94$; $p = 0.311$).

- Air relative temperature (RT) decreases by 12% in the interior, showing nonsignificant values ($H = 5.69$; $p = 0.337$).

Changes are not manifested with identical magnitude in all cases. The smaller fragments tend to register values higher or lower for the measured variables (results not shown).

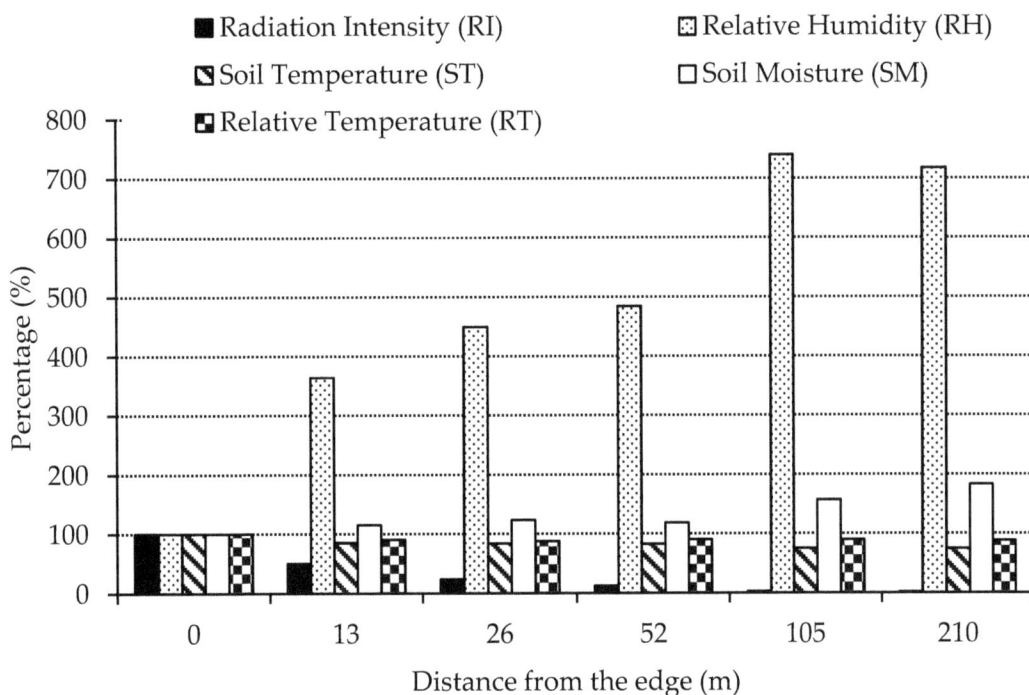

Figure 6. Microclimatic variables studied in fragments from edge to inside. Values are expressed in relative terms as a percentage of value at the edge (considered as 100%). The units are the following: RI= W/cm²; ST= °C; RT=°C; RH= %; SM= %.

Microclimatic variables are interrelated. Thus, for example, the RH and the RT are inverse and strongly related; the RI and RT relate directly and the RH and RI in reverse. This means that the intensity of radiation reaching the edge of the plot is influencing the relative temperature directly (higher radiation and higher relative temperatures) and inversely with relative humidity (greater radiation and lower relative humidity). In addition, the relative humidity and temperature inversely influence themselves (where there are higher values of relative temperature, there are lower values of relative humidity).

In the AGB case, the relative participation of each species to the biomass stock varies according to site between 9% and 22 % for *C. multiflorum*, 5% and 79% for *P. rhamnoides*, 0% and 15% for *A. urundeuva*, 11% and 48% for *A. colubrina*, and between 0% and 23% for *C. angustifolia*. In general, the best-represented species is *P. rhamnoides*, followed by *A. colubrine*, and *C. multi-florum*. The fraction of ≤10 cm dbh ("sprout") contributes to their maximum values up to 6% of total AGB per site. AGB decreases significantly ($H = 53.66$; $p < 0.001$) from site 5 (179 ± 36 t/ha) to site 1 (116.4 ± 32.2 t/ha), site 2 (106 ± 44.6 t/ha), site 4 (16 ± 6.7 t/ha), and lastly site 3 (10.37 ± 4.1 t/ha). The studied species represent approximately 86–90% of the total in the case of the forest (according to plot). In the fragments, the five studied species not only have lower AGB but also have proliferated heliophyllum species, typical of open environments, and species composition has changes (results not shown). It cannot be concluded that carbon sequestration in vegetation is less because of the microclimatic edge effect. Although there are clear differences in the AGB_{10}, the correlation of different distance values does not give significant values

($r = 0.03$; $p = 0.804$), nor in the AGB_0 ($r = 0.20$; $p = 0.134$). The AGB of key species differs among fragments, but it cannot be said that a whole biomass has declined, since other shrubs and herbaceous species have proliferated. Larger studies are necessary to evaluate this aspect in depth.

Carbon sequestration in SOC, estimated up to 10 cm depth, increases from 19.3 ± 5 tC/ha in the site 3 (small forest fragment) to 23.4 ± 5 tC/ha in the site 4 (small forest fragment), 28.8 ± 7.5 tC/ha in the site 1 (large forest fragment), 28.9 ± 12.2 tC/ha in the site 2 (large forest fragment), and 34.8 ± 8.8 tC/ha in the forest or site 5 (Figure 7).

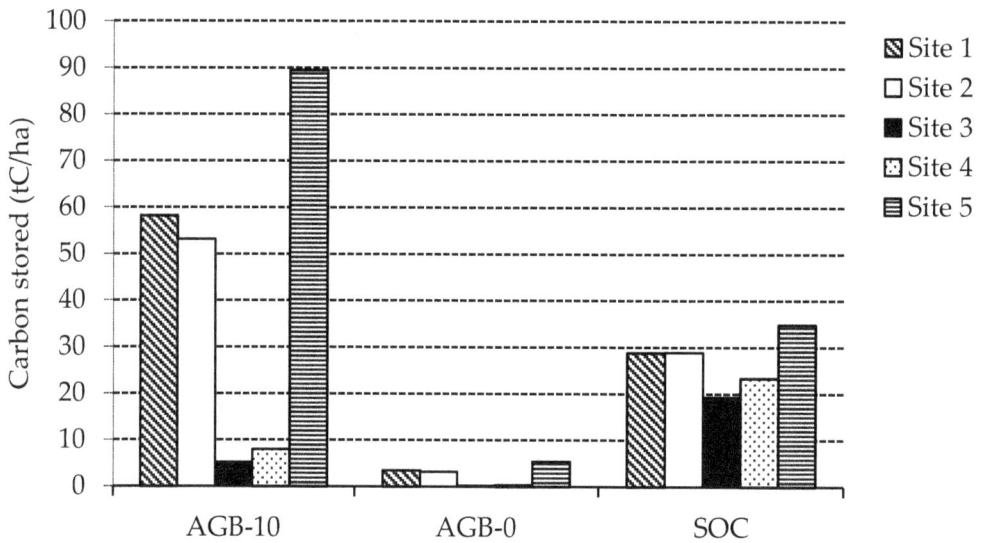

Figure 7. Carbon stock and contribution of each carbon pool studied (AGB includes only five species studied). The acronyms AGB_{10}, AGB_0 and SOC are explained in the text.

It can be assumed that the influence of these changes will affect, in the middle or long term, the composition and facilitate the establishment the different species, according to their requirements. Mainly, the dominant tree species (climax) could result in changes in its germination and survival, promoting the success of pioneers species implantation, and altering the original composition and structure of the forest [18].

4. Main remarks

The studies presented in this chapter offer insight into the varied potential of the Pedemontana Jungle for sequestration of atmospheric carbon, and how this potential can be influenced by human intervention in processes of deforestation, degradation, and fragmentation.

The carbon stock estimated for the Pedemontana Jungle ranges from 162 tC/ha (in Coronel Moldes) to 272 tC/ha (in Calilegua). In all cases, greater carbon storage occurs in the AGB fraction (from 47% to 55% of the total), where AGB_0 fraction provides between 6 and 10% of the total stock. Soil (SOC) constitutes the second most important carbon pool. Its contributions range from 28% to 39% according to the site.

The Pedemontana Jungle sequesters 43% more carbon than the Chaco forest at the same latitude. Moreover, the potential of carbon sequestration in the Pedemontana Jungle increases as the latitude decreases, sequestrating 28% more carbon at 22° than 24° south latitude.

Carbon sequestration in the Pedemontana Jungle sectors least affected by humans (degradation) is 23% higher than in more degraded areas. There are no advantages for sites that are legally protected (i.e., carbon sequestration is approximately similar). Forest degradation practices such as unsustainable timber production, overharvesting of fuel wood, extensive cattle ranching, and fires at the edge of forest fragments are less easily observed than deforestation, but they can contribute substantially to emissions. Forest degradation can also be a precursor to deforestation. These multiple changes in land use and forest area need to be monitored at the national level.

The Pedemontana Jungle sectors that have been left isolated are subject to edge effect, with changes clearly visible in microclimatic variables. The AGB in fragments is notably reduced for the main five tree species studied, but the species composition has also changed.

The potential impact of climate change on forest remnants is still unpredictable and depends on each one's resilience, on the remnant's adaptive capacity to climate change, and the magnitude and intensity of the phenomenon manifested in each area. At the same time, deforestation, degradation, and fragmentation of the Pedemontana Jungle could be affecting its ecological and social integrity, and the ability to provide ecosystem services of supply and regulation in the long term, and therefore its ability to respond to the global climate change impact.

Human management has taken over the ecosystem services that sustain the most important production systems from an economic standpoint. For example, irrigation water for cattle pastures and soil for agriculture. In many forests, such as the Pedemontana Jungle, other ecosystem services, for example, cultural or climatic regulation, are subordinated to these major objectives. The consequences of this imbalance in handling are shown negatively in the middle and long term, whereas in the short term, it cannot be seen most of the time. Vulnerable ecosystems are thus generated from the biophysical and social point of view, with a reduced capacity to respond to additional disturbances such as global climate change.

Forests require immediate support, with long-term policies independent of the ideologies, and management plans developed on technical bases, which are based on compliance with Article 41 of the National Constitution, *"All citizens enjoy the right to a healthy and balanced environment, suitable for human development and for productive activities that meet present needs without compromising those of future generations; and have the duty to preserve it..."* Land use plans should prioritize the conservation of ecosystems of high ecological value, such as the Pedemontana Jungle or Chaco, moreover, in a province where the natural biodiversity is accompanied by cultural biodiversity (with nine aboriginal ethnic groups), and where the forests are the principal sustainers of life.

Acknowledgements

Supported by the Erasmus Mundus Action 2 Programme of the European Union. Special thanks to CONICET and the Research Council of National University of Salta, both of Argentina, and National University of Salamanca, Spain. The Municipality of Coronel Moldes and Colonia Santa Rosa are gratefully acknowledged. The authors thank Sandra Brown, Milena Segura, and Angelina Martínez-Yrízar for information, suggestions, and comments. Andrés Tálamo is acknowledged for his statistical advice. Thanks to the National Parks Administration and Environment Secretary of the province of Salta, for entry permits to protected areas and logistical support. This work could not have been completed without the invaluable help of the students who assisted during field trips.

Author details

Silvina M. Manrique and Judith Franco

*Address all correspondence to: silmagda@unsa.edu.ar

Non Conventional Energy Resources Investigation Institute (INENCO) of National University of Salta (UNSa) and National Council of Scientific and Technical Research (CONICET), Salta, Argentina

References

[1] Keenan R, Reams G, Achard F, Freitas J, Grainger A, Lindquist E. Dynamics of global forest area: results from the FAO Global Forest Resources Assessment. *Forest Ecology and Management*, 2015. 352: pp. 9–20. doi.org/10.1016/j.foreco.2015.06.014

[2] Aerts R, Honnay O. Forest restoration, biodiversity and ecosystem functioning. *BMC Ecology*, 2011. pp. 11–29. doi: 10.1186/1472-6785-11-29

[3] Bonan GD. Forest and climate change: forcing, feedbacks and the climate benefits of forest. *Science*, 2008. 320: pp. 1444–1449.

[4] Nelson E, Mendoza G, Regetz J, Polasky S, Tallis H, Cameron D, Chan KM, et al. Modeling multiple ecosystem services, biodiversity conservation, commodity production and tradeoffs at landscape scales. *Frontiers in Ecology and Environment*, 2009. 7: pp. 4–11. doi: 10.1890/080023

[5] Gibson L, Lee T, Koh L, Brook B, Gardner T, Barlow J, Peres C, et al. Primary forests are irreplaceable for sustaining tropical biodiversity. *Nature*, 2011. 478(7369): pp. 378–381. doi: 10.1038/nature10425.

[6] Hansen M, Potapov P, Moore R, Hancher M, Turubanova S, Tyukavina A, Thau D, et al. High-resolution global maps of 21 st-century forest cover change. *Science*, 2013. 342(6160): pp. 850–853. doi: 10.1126/science.1244693

[7] Pinto S, Mendes G, Santos A, Dantas M, Tabarelli M, Melo F. Landscape attributes drive complex spatial microclimate configuration of Brazilian Atlantic forest fragments. *Tropical Conservation Science*, 2010. 3(4): pp. 399–402.

[8] Miki T, Ushio M, Fukui S, Kondoh M. Functional diversity of microbial decomposers facilitates plant coexistence in a plant-microbe-soil feedback model. *Proceedings of the National Academy of Sciences of the United States of America*, 2010. 107: pp. 14251–14256. doi: 10.1073/pnas.0914281107.

[9] Van der Putten W, Bardget R, Bever J, Bezemer T, Casper B, Fukami T, Kardol P, et al. Plant–soil feedbacks: the past, the present and future challenges. *Journal of Ecology*, 2013. 101(2): pp. 265–276. doi: 10.1111/1365-2745.12054.

[10] DeFries R, Achard F, Brown S, Herold M, Murdiyarso D, Schlamadinger B, Souza C. Earth observations for estimating greenhouse gas emissions from deforestation in developing countries. *Environmental Science and Policy*, 2007. 10: pp. 385–394.

[11] Dixon R, Brown S, Houghton R, Solomon A, Trexier M, Wisniewski J. Carbon pools and flux of global forest ecosystems. *Science*, 1994. 263(14): pp. 185–190.

[12] Pan Y, Birdsey R, Fang J, Houghton R, Kauppi P, Kurz W, Phillips O, et al. A large and persist carbon sink in the world's forest. *Science*, 2011. 333: pp. 988–993. doi: 10.1126/science.1201609

[13] Malhi Y, Roberts T, Betts R, Killeen T, Li W, Nobre C. Climate change, deforestation and the fate of the Amazon. *Science*, 2008. 319(169): pp. 169–172. doi: 10.1123/science.1146961

[14] Bala G, Caldeira K, Wickett M, Phillips T, Lobell D, Delires C, Mirin A. Combined climate and carbon-cycle effects of large-scale deforestation. *Proceedings of the National Academy of Sciences of the United States of America*, 2007. 104(16): pp. 6550–6555. doi: 10.1073/pnas.0608998104

[15] Betts R, Cox P, Collins M, Harris P, Huntingford C, Jones D. The role of ecosystem-atmosphere interactions in simulated Amazonian precipitation decrease and forest dieback under global climate warming. *Theoretical and Applied Climatology*, 2004. 78(1): pp. 157–175. doi: 10.1007/s00704-004-0050-y

[16] Intergovernmental Panel Climate Change (IPCC) (eds.). *Climate change 2013. Fifth Assessment Report (AR5)*. NY, USA: Cambridge University Press. 2014. 32p.

[17] Ryan C, Berry N, Joshi N. Quantifying the causes of deforestation and degradation and creating transparent REDD baselines: a method and case study from central Mozambique. *Applied Geography*, 2014. 53: pp. 45–54. doi: 10.1016/j.apgeog.2014.05.014

[18] Newman B, Ladd P, Brundrett M, Dixon K. Effects of habitat fragmentation on plant reproductive success and population viability at the landscape and habitat scale. *Biological Conservation*, 2013. 159: pp. 16–23. doi: 10.1016/j.biocon.2012.10.009

[19] Metzger M, Rounsevell M, Acosta-Michlik L, Leemans R, Schroter. The vulnerability of ecosystem services to land use change. *Agriculture, Ecosystems and Environment*, 2006. 114: pp. 69–85. doi: 10.1016/j.agee.2005.11.025

[20] Volante J, Alcaraz-Segura D, Mosciaro M, Viglizzo E, Paruelo J. Assessing the effect on land clearing on ecosystem services provision in north-western Argentina. *Agriculture, Ecosystems and Environment*, 2012. 154: pp. 12–22.

[21] Viglizzo E, Jobbágy E (eds.). *Expansión de la frontera agropecuaria en Argentina y su impacto ecológico-ambiental*. Buenos Aires, Argentina: INTA. 2010. 106p. ISBN N° 978-987-1623-83-9.

[22] Barros V, Vera C (eds.). *Cambio climático en Argentina: tendencias y proyecciones. Tercera Comunicación Nacional de la República Argentina a la Convención Marco de las Naciones Unidas sobre Cambio Climático*. Buenos Aires, Argentina: Secretaría de Ambiente y Desarrollo Sustentable de Argentina (SAyDS). 2014.

[23] Secretaría de Ambiente y Desarrollo Sustentable de Argentina (SAyDS) (eds.). *Monitoreo de la superficie de bosque nativo de Argentina. Periodo 2010-2013*. Buenos Aires: SAyDS. 2014. 50p.

[24] Cabrera AL. *Regiones fitogeográficas de Argentina. Enciclopedia Argentina de Agricultura y Jardinería*. Buenos Aires: Acme Saci. 1994. 45p.

[25] Brown A, Blendinger P, Lomáscolo T, Bes P. *Selva Pedemontana de las Yungas. Historia natural, ecología y manejo de un ecosistema en peligro*. Yerba Buena, Tucumán: Ediciones del subtrópico. 2009. 490 p. ISBN: 978-987-23533-5-3

[26] Secretaría de Ambiente y Desarrollo Sustentable de Argentina (SAyDS) (eds.). *Los bosques nativos argentinos. Un bien social*. Buenos Aires: SAyDS. 2007.

[27] Blendinger P, Rivera L, Alvarez M, Nicolossi G, Politi N. Selección de áreas prioritarias para la conservación de las aves de la Selva Pedemontana de Argentina y Bolivia, in: Brown A, Blendinger P, Lomáscolo T, Bes P (eds.). *Selva Pedemontana de las Yungas. Historia natural, ecología y manejo de un ecosistema en peligro*. Yerba Buena, Tucumán: Ediciones del Subtrópico. 2009. 409–415.

[28] Organización de las Naciones Unidas para la Alimentación y la Agricultura (FAO). *Base referencial mundial del recurso suelo*. Roma, Italia: FAO. 2007.

[29] Montenegro C, Strada M, Bono J, Gasparri N, Manghi E, Parmuchi M, Brouver M (eds.). *Estimación de la pérdida de superficie de bosque nativo y tasa de deforestación en el norte de Argentina*. Buenos Aires: SAyDS. 2005.

[30] Macdicken K. *A guide to monitoring carbon storage in forestry and agroforestry projects*. USA: Winrock International Institute for Agricultural Development. 2007. 91 p.

[31] Cairns M, Brown S, Helmer E, Baumgardner G. Root biomass allocation in the worlds upland forests. *Oecologia*, 1997. 111: pp. 1–11.

[32] Brown S. Estimating biomass and biomass change of tropical forests. *FAO Montes*, 1997. 134: pp. 1–55.

[33] Chave J, Andalo C, Brown S, Cairns M, Chambers J, Eamus D et al. Tree allometry and improved estimation of carbon stocks and balance in tropical forests. *Oecologia*, 2005. 145: pp. 87–89.

[34] Gehring C, Park S, Denicha M. Liana allometric biomass equations for Amazonian primary and secondary forest. *Forest ecology and management*, 2004. 195: pp. 69–83. doi: 10.1016/j.foreco.2004.02.054

[35] Sevola Y. Cubicación de árboles en el Inventario Forestal del Noroeste Argentino. *FAO Montes*, 1975. 20: pp. 1–42.

[36] INTI-CITEMA. Listado de densidades secas de maderas. *Series Técnicas Instituto Nacional de Tecnología Industrial y Centro de Investigación Tecnológico de la madera*, 2007. pp. 1–8.

[37] Bianchi A, Yañez C. *Las precipitaciones en el noroeste argentino*. Salta, Argentina: INTA-EEA Salta. 1992.

[38] Tortorelli L. *Maderas y bosques argentinos*. Buenos Aires: Secretaría de Ambiente y Desarrollo Sustentable de Argentina. 1956.

[39] Brown S, Gillespie A, Lugo A. Biomass estimation methods for tropical forests with applications to forest inventory data. *Forest Science*, 1989. 35(4): pp. 381–902.

[40] Ripley S, Krzic M, Bradfield G, Bomke A. Land use impacts on selected soil properties of the Yungas/Chaco transition forest of Jujuy province, northwestern Argentina: a preliminary study. *Canadian Journal of Soil Science*, 2010. 90: pp. 679–683. doi: 10.4141/CJSS09101

[41] Bonino E. Changes in carbon pools associated with land-use gradient in the Dry Chaco. *Forest Ecology and Management*, 2006. 223: pp. 183–189. doi: 10.1016/j.foreco. 2005.10.069

[42] Manrique, S.; Franco, J.; Núñez, V. y Seghezzo, L. (2011). *Potential of native forests for the mitigation of greenhouse gases in Salta, Argentina. Biomass and Bioenergy* 35(5): 2184-2193. ISSN: 0961-9534. DOI: 10.1016/j.biombioe.2011.02.029.

[43] Rodrigues A, Andelman S, Bakarr M, Boitani L, Brokks T, Cowling M, et al. Effectiveness of the global protected area network in representing species diversity. *Nature*, 2004. 428: pp. 640–665. doi: 10.1038/nature02422

[44] Tittensor D, Walpole M, Hill S, Boyce D, Britten G, Burgess N, Visconti P. A mid-term analysis of progress toward international biodiversity targets. *Science*, 2014. 346(6206): pp. 241–244. doi: 10.1126/science.1257484

8

Energy for Sustainable Development: The Energy–Poverty–Climate Nexus

Melanie L. Sattler

Additional information is available at the end of the chapter

Abstract

Worldwide, 1.4 billion people lack access to electricity, and 2.7 billion people rely on traditional biomass for cooking. Most people living in energy poverty—without electricity access and/or using traditional biomass for cooking—are from rural areas of Sub-Saharan Africa, India, and other developing Asian countries (excluding China). At the same time, the poorest people are the most likely to suffer from the impacts of climate change.

Fortunately, innovative, sustainable energy technologies can allow developing countries to leapfrog to low-carbon renewable energy, while at the same time alleviating extreme poverty. Increasing energy access, alleviating rural poverty, and reducing greenhouse gas emissions can thus be complementary, their overlap defining an energy–poverty–climate nexus. Transitioning to more efficient low-carbon energy systems in rural areas can generate greater returns than similar efforts in industrialized areas.

Accordingly, this chapter provides an overview of: a) The linked problems facing developing countries of energy access, poverty, and climate change, and how these problems interact and compound each other. b) Potential renewable energy solutions, including off-grid solar, wind, clean biomass, micro-hydro, and hybrid systems. For each energy option, benefits and challenges will be discussed, along with examples of successful small-scale use in rural areas of developing countries.

Keywords: Sustainable energy, climate change, developing countries, renewable energy, energy–climate–poverty nexus

1. Introduction

Most of the world's people without access to electricity or clean energy are from rural areas of developing countries. At the same time, a person in a developing country is 79 times more likely to suffer from a climate-related disaster than a person in a developed country, according to the United Nations Development Programme [1].

Fortunately, for developing countries, increasing energy access, alleviating poverty, and addressing climate change can all be accomplished via sustainable energy. In fact, implementing renewable energy in rural areas can generate greater returns in terms of reduced greenhouse gas emissions than similar efforts in industrialized areas. Accordingly, this chapter will provide an overview of the current energy access problem in developing countries and potential innovative sustainable solutions. Off-grid renewable electricity options to be discussed include solar, wind, clean biomass, micro-hydro, and hybrid systems.

2. The Problems

The problems of energy access, poverty, and climate change are intertwined in the developing world. The poor often lack access to energy at all or have access only to inefficient and unhealthy forms of energy. Lack of energy makes it more difficult to address other aspects of poverty, such as lack of education or health care (imagine operating a school or hospital without access to electricity). As the poor gain access to energy, their contribution to climate change will increase, unless they leapfrog to renewable energy technologies. Unfortunately, the poor are the most vulnerable to many impacts of climate change, including increased food insecurity and amplified health risks. These interrelationships that constitute the energy–poverty–climate nexus problem are discussed in more detail below.

2.1. The Energy–Poverty Nexus

Worldwide, 1.4 billion people (20% of the world's population) currently lack access to electricity, and 2.7 billion people (40% of the world's population) rely on inefficient and unhealthy forms of biomass [2]. Air pollution levels indoors from 3-stone fires or inefficient stoves using biomass are many times higher than typical outdoor levels, even those in highly polluted cities. The World Health Organization (WHO) estimates that over 1.5 million die prematurely each year from household air pollution due to inefficient biomass combustion [2]. These deaths from cancer, respiratory infections, and lung diseases account for 4% of the global burden of disease—more deaths than those from malaria (1.2 million) or tuberculosis (1.6 million) [1]. Many of those who die are young children, who spend hours each day breathing in smoke from the cookstove [2].

Moreover, in regions where households rely heavily on unhealthy forms of biomass, women and children are typically responsible for fuel collection—a time-consuming and exhausting task. This strenuous work without sufficient recuperation can cause serious long-term physical damage for women. Heavy reliance on biomass can also cause land degradation, including deforestation, and local and regional air pollution [2].

As shown in Table 1, most of the people living in energy poverty—without electricity access and/or using traditional biomass for cooking—are from rural areas of developing countries, the majority in Sub-Saharan Africa, India, and other developing Asian countries (excluding China) [2]. At current growth rates, about half a billion "energy poor" will be added over the next 20 years [1].

Region	Number of people lacking access to electricity (millions)	Number of people relying on the traditional use of biomass for cooking (millions)
Africa	587	657
Sub-Saharan Africa	585	653
Developing Asia	799	1937
China	8	423
India	404	855
Other Asia	387	659
Latin America	31	85
Developing countries*	1438	2679
World**	1441	2679

*Includes Middle East countries.

**Includes Organisation for Economic Co-operation and Development (OECD) and transition economies.

Table 1. Number of people lacking access to electricity and relying on the traditional use of biomass for cooking, 2009 (million) [2]

Improving access to cost-effective, sustainable energy technologies is critical for addressing poverty in developing countries [1]. Although improving energy access is not one of the 8 globally agreed Millennium Development Goals (MDGs), it is a cross-cutting issue that directly impacts achievement of the goals [1]. More and better energy services are needed to end poverty, hunger, educational disparity between boys and girls, the marginalization of women, major disease and health service deficits, as well as environmental degradation [1, 2]. Without modern energy services, basic social goods such as health care and education are more costly in both real and human terms, and economic development is harder to perpetuate [1]. A clear correlation exists between energy and the Human Development Index (HDI) [3]. As the International Energy Agency and United Nations state, "Access to modern forms of energy is essential for the provision of clean water, sanitation and healthcare and provides great benefits to development through the provision of reliable and efficient lighting, heating, cooking, mechanical power, transport and telecommunication services" [2].

For the 1.4 billion people that lack access to electricity, they either live in locations too remote to be connected or cannot afford the fee to connect [4]. For locations that are off-grid, fossil fuels are often unaffordable due to the cost of delivery to remote locations [5].

2.2. The Energy–Climate Nexus

The climate change problem is largely a fossil fuel problem. According to the Intergovernmental Panel on Climate Change (IPCC) 2007 report, at least 57% of greenhouse gas emissions globally stem from burning of fossil fuels [6]. For carbon dioxide (CO_2), the most important anthropogenic greenhouse gas, 74% of emissions are due to combustion of fossil fuels [6]. When fossil fuels are burned for energy, the carbon stored in them, originally from biomass such as algae, is emitted almost entirely as CO_2. These fossil fuels include coal, oil, and natural gas, which are burned in electric power plants, automobiles, industrial facilities, and other sources.

2.3. The Poverty–Climate Nexus

The poorest people in the world are the most likely to suffer from impacts of climate change. According to the United Nations Development Programme (UNDP), a person in a developing country is 79 times more likely to suffer from a climate-related disaster than a person in a developed country [2]. The poor are especially vulnerable to the impacts of climate change, including reduced agricultural productivity and increased food insecurity; heightened water stress and insecurity; rising sea levels and increased exposure to climate disasters; loss of ecosystems and biodiversity; and amplified health risks [1]. According to the Human Development Report (HDR) 2007–2008, failure to address climate change will consign and trap the poorest 40% of the world's population, some 2.6 billion people, in downward spirals of deprivation [7]. Providing energy access will help poor areas adapt in the face of a changing climate [8].

Reductions in greenhouse gas emissions must include both developed and developing countries, including those with significant numbers of people living in poverty. According to the U.S. Environmental Protection Agency (EPA), greenhouse gas emissions from developing countries are expected to exceed those from developed countries in 2015 [9]. According to Casillas and Kammen [8], every dollar spent on the transition to more efficient low-carbon energy systems in rural areas has the potential to produce greater carbon mitigation returns than in more industrialized areas.

3. Potential Solutions

Since the problems of energy access, poverty, and climate change are interrelated in developing countries, solutions can be designed to solve all 3 problems simultaneously. Innovative sustainable energy technologies can allow developing countries to leapfrog to low-carbon renewable energy, while at the same time alleviating extreme poverty. Increasing energy access, alleviating poverty, and reducing greenhouse gas emissions can thus be complementary, their overlap defining an energy–poverty–climate nexus solution.

As mentioned above, according to the article published by Casillas and Kammen [8] recently in *Science*, transitioning to more efficient low-carbon energy systems in rural areas can generate greater returns than similar efforts in industrialized areas. Urban cities of developing countries may have access to electricity for lighting. Rural areas typically lack access altogether; hence, the need in these areas is the greatest [9]. Access to electricity in rural areas, even at modest consumption levels, can dramatically improve a community's quality of life. For example, electric lamps can allow children to study at night, and radios and cellular phones can greatly improve communication pathways [10]. This section will accordingly focus on renewable energy solutions for rural areas of developing countries.

Centralized electrification requires massive amounts of capital [10]. The dispersed nature of houses and low potential demand create little incentive for power companies to provide access to rural areas. In addition, extending the grid may be unrealistic due to transmission line costs or hard terrain [5]. Thus, in rural areas, off-grid and mini-grid solutions make the most sense. Such systems can consist of a single home or several small homes and businesses. The systems can be incremental and scalable and applied to many different conditions and environments [10]. Off-grid and mini-grid options for renewable electricity include solar, wind, clean biomass, and micro-hydro. These options for renewable power will be discussed in more detail below.

3.1. Solar Power

Solar energy is abundant in many locations in the developing world [5]. Many regard it as the most promising renewable source for developing countries [5, 11, 12]. The use of solar energy produces no on-site air pollutants [although pollutants are typically generated in the process of manufacturing photovoltaics (PV) cells].

Solar energy can be utilized in two ways: direct heat energy for various purposes (heating water, heating space) and direct current electricity generation using PV system. Electrical energy can be used immediately to pump water for irrigation or for refrigeration, lighting, or other purposes; alternatively, it can be stored in a rechargeable battery for later use [5]. This can help solve problems associated with its intermittency [12].

Individual houses can have their own PV system for lighting and small appliances, such as radio and mobile phone charging. A village can benefit from a larger PV system, with a micro-grid structure.

3.1.1. Solar Challenges

The primary barrier to widespread implementation of the solar PV technology is its cost, due to the high cost of the silicon base material and associated manufacturing processes [1]. In addition, production of solar cells currently requires sophisticated and expensive manufacturing facilities and highly trained personnel, which may not be available in developing countries. Nicole Kuepper, a Ph.D. student, won the Eureka Prize for Young Leaders in Environmental Issues and Climate Change for developing a simple, inexpensive way of producing solar cells in a pizza oven. The process uses a low-cost inkjet printing process, aluminum spray, and a low-temperature pizza oven, meaning that the solar cells can be made without high-tech environments or high-cost inputs [1].

Solar water heaters (SWHs) are relatively expensive to install ($500–$2100), although the initial investment can be recovered through future electricity savings [1]. For many families in rural areas, the purchase of a solar lighting set, even for lighting service only, is so hard that a systematic approach has to be designed, which enables them to pay only for the lighting service received instead of owning the whole hardware [5].

3.1.2. Solar Success Stories

Solar lanterns. Typically, solar-powered lanterns use solar energy to charge a battery that powers a solid-state light-emitting diode (LED), the most efficient lighting technology on the market. One solar lantern, the Mighty Light, costs around US$ 45 and lasts up to 30 years. It has replaced polluting dangerous kerosene lamps for thousands of households in Afghanistan, Guatemala, India, Pakistan, and Rwanda [1]. As of November 2010, around 9000 versions of a solar lantern called the "solar tuki" had been installed in Nepal [1].

Solar home systems. A solar home system is a PV system with capacity of 10–40 Wp (peak Watts). By 2007, Grameen Shakti had installed 100,000 solar home systems to power lights, motors, pumps, televisions, mobile phones, and computers in Bangladesh [1].

SELCO of India is an organization successfully installing solar home and business systems to provide electricity for lighting, SWHs, solar inverter systems (for use in communications and computing), and small business appliances. Since 1995, SELCO has been providing solar energy solutions to underserved households and businesses in India, based on the ideas that poor people can afford sustainable technologies, poor people can maintain sustainable technologies, and social ventures can be run as commercial entities [1].

Community solar systems. Community solar PV systems are commonly used for pumping water for drinking and irrigation. The solar panel may vary between 130 Wp and 40 kWp. The Promethean Power project promotes a version of community solar systems using concentrated solar thermal power rather than PV. The system, which can be manufactured locally, concentrates solar thermal energy to heat a fluid refrigerant. The solar thermal system is combined with a unique microscale generator adapted and scaled to suit the needs of underserved communities. The heated fluid expands through a rotary vane turbine (an automobile power-steering pump) to make mechanical energy that spins a generator (an automobile alternator). Massachusetts Institute of Technology (MIT) is installing the systems in Lesotho in Africa [1].

3.2. Wind Power

Like solar, the fuel source for wind power is free and unlimited. The use of wind turbines to generate electricity via a generator produces no on-site air pollution, although a small amount of emissions is produced during manufacture of the turbines. One life-cycle assessment found that off-grid wind turbines reduce greenhouse gas emissions by 93%, compared to off-grid diesel power generation systems [13]. An additional advantage is that wind turbines are simple mechanical systems that can be easily maintained and repaired [10].

3.2.1. Wind Challenges

One of the challenges associated with wind power is its intermittency. Researchers are developing solutions to this problem for off-grid wind systems. Short term, the electrical energy generated by the turbine can be stored in a battery [14]. Researchers have developed controllers that maximize capture of wind energy and avoid battery overcharge [15, 16]. Other researchers have proposed a hybrid energy storage system that can provide uninterrupted power, according to simulations. In the hybrid storage system, a battery is used for short-term energy storage, and a water electrolysis hydrogen system is used for long-term energy storage due to hydrogen's high mass energy density and very low leakage [17].

Another challenge is selecting an appropriate location for the turbine, due to the highly localized nature of wind. Low-cost anemometers may help alleviate this problem, but time must be spent to collect a sufficient amount of data [10]. Areas particularly suited to wind power because of their typical high wind velocities include coastlines, high ground, and mountain passes [12]. Wind power does not need water, so it is suitable for dry areas.

A third challenge is the need for a tower, so that the turbine is at least 10 m above the nearest obstacle. The tower itself could cost more than the wind turbine. Researchers are examining towers made from bamboo and other common and low-cost materials [10].

3.2.2. Wind Success Stories

The cost of commercially available wind turbines is several thousand US dollars per kilowatt, which is out of reach for most rural residents of developing countries [10]. Low-cost wind turbines with timber blades have been demonstrated successfully in Nepal [18]. Low-technology wind turbine generators, which can be made by people with limited technical skills, no advanced machining equipment, low capital cost, and limited exotic materials, have also been demonstrated [10]. The low-technology wind turbine was constructed in a joint effort by the IEEE Power & Energy Society Community Solutions Initiative and the Puget Sound Professional Chapter of Engineers Without Borders, USA.

3.3. Clean Biomass

Biomass sources in rural areas include human excrement, animal manure, and agricultural wastes. Biomass can be burned directly to produce heat energy or electricity via a microturbine, or it can be degraded by anaerobic microbes to produce biogas. Although burning biomass in inefficient cookstoves contributes to illness via indoor air pollution, as described above, biomass can be burned using clean technologies or used to generate clean-burning biogas. Biogas, typically 60–70% methane, can be used directly in natural gas-powered appliances or burned to generate electricity via a microturbine and generator [19]. It is expected that microturbines powered by biogas might eventually be competitive with diesel engines for village-scale power applications, with relatively low maintenance costs, high reliability, long lifetimes, and low capital costs [1]. Fuel cells might ultimately prove able to generate power at village scales from biogas, at very high efficiencies [1].

Anaerobic processes that produce methane from waste solve 2 problems at once: waste and energy. Anaerobic processes provide some of the simplest and most practical methods for minimizing public health hazards from human and animal wastes. Pathogens such as schistosome eggs, hookworm, flat/tape worm, dysentery *Bacillus*, poliovirus, *Salmonella*, and

Bacillus for paratyphoid are destroyed. A residence time of 14 days at >35°C in a small-scale system in a developing country can provide 99+% removal of pathogens, with the exception of roundworm [20, 21].

In addition, the solid residue from anaerobic waste treatment processes is a valuable fertilizer, which is stabilized and almost odorless. This fertilizer is especially a benefit in developing countries, due to its potential to boost crop yields.

3.3.1. Biomass Challenges

Ideally, energy should be produced from biomass that is not edible and that cannot be grown in places where edible crops could be grown, so that competition between uses of crops for energy and food does not become an issue. Producing energy from wastes avoids this issue.

Because of limits on the amount of land accordingly available for growing plants that can be used for energy, bioenergy cannot be viewed globally as the sole replacement or substitute for fossil fuels, but rather as one element in a broader portfolio of renewable energy sources [1]. In rural locations in developing countries without current access to electricity, however, biomass can provide a transformative local power source.

3.3.2. Biomass Success Stories

An improved biomass cookstove designed by Prakti Design Lab to meet the cooking requirements of rural households is around 40% percent more fuel efficient than traditional cookstoves and emits 70–80% less smoke [1].

In Senegal, proliferating invasive aquatic plants are being transformed into combustible pellets that can be used for cooking, replacing wood and charcoal. By impacting lake water quality, the plants' proliferation caused an increase in waterborne diseases. The plants are also created problems for fishermen, by jamming their nets, and farmers, by reducing access to water for livestock. Local fishermen and farmers will be recruited as plant removers, and 20 additional local workers will be hired and trained to manage the pellet production process. Based on capacity production of the compaction machine (4,000 kg/week) and a local price of US$ 0.28/kg, the pellets could generate income of about US$ 1,120/week for the local population [1].

3.4. Micro-Hydro

Micro-hydro systems use the natural flow of water to yield up to 100 kW output of electrical energy [22]. Simplicity, efficiency, longevity, reliability, and low maintenance costs make these systems attractive for rural development [23]. Like solar and wind, the fuel source for micro-hydro power is free, and the use of hydro-powered turbines to generate electricity produces no on-site air pollution.

Unlike large hydroelectric plants, micro-hydro systems do not require a dam and reservoir, which minimize their environmental damage. A portion of the river's flow is diverted to the micro-hydro intake. A settling tank may be used to allow silt to settle out of the water. A screen or bars screen out floating debris and fish. The water then flows through a channel, pipeline, or pressurized pipeline (penstock) to the powerhouse, which houses a turbine or waterwheel. The turbine turns a generator to produce electricity [22]. A variety of turbines may be used, including a Pelton wheel for high head, low-flow water supply, or a propeller-type turbine for low-head installations [24, 25].

3.4.1. Micro-Hydro Challenges

Micro-hydro systems obviously are limited to locations with a stream or river. The flow volume must be sufficient to supply local energy needs. In addition, a sufficient quantity of falling water must be available, which usually means that hilly or mountainous sites are best. A drop of water elevation of at least 2 ft is required or the system may not be feasible; the water does not "fall" enough distance to produce enough head [22, 24]. Another limitation is that the distance from the stream or river to the site in need of energy may be considerable [23].

The power produced may fluctuate depending on how much water is flowing in the stream or river and the velocity of flow [23]. Energy can be stored in batteries, so additional reserve energy available for time of low generation and/or high demand. However, because hydro-power resources tend to be more seasonal than wind or solar resources, batteries may not be able to provide enough energy storage for summer or other seasons with severely limited water flow. Integrating the hydropower with a hybrid wind or solar system can help in areas where water flow is highly seasonal.

3.4.2. Micro-Hydro Success Stories

The main micro-hydro programs in developing countries are in mountainous regions, such as Nepal (around 2,000 installations, including both mechanical and electrical power generation) and other Himalayan countries [25]. In South America, micro-hydro programs are located in countries along the Andes, such as Peru and Bolivia. Smaller programs have been initiated in hilly areas of Sri Lanka, the Philippines, China, and elsewhere [25]. In a variety of locations, micro-hydro systems have been shown to increase employment opportunities in rural areas, which encourage young people to stay in the villages rather than drifting to the cities [25].

Maher et al. [26] describe the successful implementation of pico hydro (<5 kW) systems in two communities in Kenya. Costs for these systems were considerably less than comparable PV or auto battery systems. The systems were constructed locally using available materials and community labor.

3.5. Hybrid Systems

According to a report issued in 2010 by the International Energy Agency, UNDP, and United Nations Industrial Development Organization, combining solar, wind, biomass, and mini-hydro into an integrated/hybrid system supplying a mini-grid is probably the most promising approach to rural electrification [2]. A combination of technologies in an integrated system can promote reliability. A small backup generator may be operated on diesel, biogas, or biodiesel [5]. Hybrid village electrification systems have been implemented in various countries, including China, India, Ghana, South Africa, and Tanzania [1]. A number of studies have examined the feasibility of various kinds of hybrid off-grid systems: wind–diesel [14, 27], wind–solar [28, 30], wind–PV–diesel [31–33], hydro–PV–wind [34], wind–hydrogen [35], and solar–wind–biomass–hydro [36].

4. What Now? Next Steps

At the global level, a new development paradigm—a pro-poor global climate change agenda—should be embraced. National climate change adaptation and mitigation strategies should be directly linked with poverty reduction and sustainable development goals [1].

To ensure that every person in the world benefits from access to electricity and clean cooking facilities by 2030, the International Energy Agency, UNDP, and United Nations Industrial Development Organization estimate that investment of $36 billion per year will be required. To meet the more ambitious target of achieving universal access to modern energy services by 2030, an additional cumulative investment of $756 billion, or $36 billion per year, is needed. Although this sounds like a large number, it represents only 0.06% of average annual global gross domestic product (GDP) over the period. The resulting increase in primary energy demand and CO_2 emissions would be modest: in 2030, global electricity generation would be 2.9% higher, and CO_2 emissions would be only 0.8% higher [2]. Given that the up-front cost of new energy technologies is prohibitively expensive for poor communities, targeted financing and incentives are needed so that low-income communities, households, and entrepreneurs can invest in new energy technologies.

Students and young entrepreneurs in collaboration with nongovernmental organizations (NGOs) have done some of the most innovative work in new low-cost sustainable energy applications. Such partnerships should be promoted. The World Bank's Development Marketplace Grants, for example, provide global recognition and seed funding for creative ideas, technologies, and services that matter for development, so that they may grow and replicate.

5. Summary

The problems of energy access, poverty, and climate change are intertwined in the developing world. The poor often lack access to energy at all or have access only to inefficient and unhealthy forms of energy. As the poor gain access to energy, their contribution to climate change will increase, unless they leapfrog to renewable energy technologies. Unfortunately, the poor are the most vulnerable to many impacts of climate change, including increased food insecurity and amplified health risks. Access to energy can reduce their vulnerability to climate change impacts.

Fortunately, increasing energy access, alleviating rural poverty, and reducing greenhouse gas emissions can all be complementary, their overlap defining an energy–poverty–climate nexus.

Solar, wind, biomass, and micro-hydro systems have all been used successfully in various locations to provide off-grid renewable power to rural areas. Each has advantages and drawbacks, depending on the particular location. Combining solar, wind, biomass, and mini-hydro into an integrated/hybrid system supplying a mini-grid is probably the most promising approach to rural electrification.

Providing universal access to modern energy services by 2030 would cost only 0.06% of average annual global GDP during the period. What else could be a more worthwhile investment?

Author details

Melanie L. Sattler*

Address all correspondence to: sattler@uta.edu

University of Texas at Arlington, Arlington, TX, USA

References

[1] Cherian A. Bridging the Divide between Poverty Reduction and Climate Change through Sustainable and Innovative Energy Technologies; 2009.

[2] International Energy Agency, United Nations Development Programme, and United Nations Industrial Development Organization. Energy Poverty: How to Make Modern Energy Access Universal? Special Early Excerpt of the World Energy Outlook 2010 for the UN General Assembly on the Millennium Development Goals; 2010.

[3] Sopian K, Ali B, Asim N. Strategies for renewable energy applications in the organization of Islamic conference (OIC) countries. Renewable and Sustainable Energy Reviews. 2011; 15(9): 4706–4725. DOI:10. 1016/j. rser. 2011. 07. 081.

[4] Stapleton GJ. Successful implementation of renewable energy technologies in developing countries. Desalination. 2009; 248(1–3): 595–602. DOI:10. 1016/j. desal. 2008. 05. 107.

[5] Erbato TT, Hartkopf T. Development of renewable energy and sustainability for off-grid rural communities of developing countries and energy efficiency. In: Asia-Pacific Power and Energy Engineering Conference Proceedings (APPEEC) 2011; 25–28 March 2011; Wuhan: IEEE; pp. 1–4. DOI:10. 1109/APPEEC. 2011. 5749112.

[6] Intergovernmental Panel on Climate Change (IPCC). Fourth Assessment Report: Climate Change [Internet]. 2007. Available from: www. ipcc. ch/publications_and_data/ publications_and_data_reports. shtml#1 [Accessed 2012-02].

[7] United Nations Development Programme. Human Development Report 2007/2008— Fighting Climate Change: Human Solidarity in a Divided World. UNDP: 2008, p. 186.

[8] Casillas CE, Kammen DM. The energy-climate-poverty nexus. Science. 2010; 330(6008): 1181–1182. DOI:10. 1126/science. 1197412.

[9] US Environmental Protection Agency (US EPA). Global Greenhouse Gas Data, Figure 3 [Internet]. 2012. Available from: www. epa. gov/climatechange/emissions/globalghg. html [Accessed: 2012-02].

[10] Louie H. Experiences in the Construction of Open Source Low Technology Off-Grid Wind Turbines. IEEE Power and Energy Society General Meeting: The Electrification of Transportation and the Grid of the Future, 2011; 24–29 July 2011; San Diego, CA: IEEE; pp. 1–7. DOI:10. 1109/PES. 2011. 6038924.

[11] Urmee T, Harries D, Schlapfer A. Issues related to rural electrification using renewable energy in developing countries of Asia and Pacific. Renewable Energy. 2009; 34(2): 354–357. DOI:10. 1016/j. renene. 2008. 05. 004.

[12] Xia X, Xia J. Evaluation of potential for developing renewable sources of energy to facilitate development in developing countries. In: Asia-Pacific Power and Energy Engineering Conference Proceedings (APPEEC), 2010; 28–31 March 2010; Chengdu: IEEE; pp. 1–3. DOI:10. 1109/APPEEC. 2010. 5449477.

[13] Fleck B, Huot M. Comparative life-cycle assessment of a small wind turbine for residential off-grid use. Renewable Energy. 2009; 34 (12): 2688–2696. DOI:10. 1016/j. renene. 2009. 06. 016.

[14] Shaahid SM, El-Amin I, Rehman S, Al-Shehri A, Ahmad F, Bakashwain J. Dissemination of off-grid hybrid wind-diesel-battery power systems for electrification of isolated settlements of hot regions. International Journal of Sustainable Energy. 2007; 26(2): 91–105. DOI:10. 1080/14786450701549873.

[15] Owais M, Aftab MS. An off-grid model setup for wind electric conversion system. In: IEEE Region 10 Annual International Conference, Proceedings/TENCON; 23–26 January 2009; Singapore: IEEE; pp. 1–5. DOI:10. 1109/TENCON. 2009. 5395834.

[16] Eren S, Hui JCY, Yazdani D. A high performance wind-electric battery charging system. In: Canadian Conference on Electrical and Computer Engineering (CCECE '06); May 2006; Ottawa, Ontario: IEEE; pp. 2275–2277. DOI:10. 1109/CCECE. 2006. 277806.

[17] Zhou K; Lu W. Self-sustainable off-grid wind power generation systems with hybrid energy storage. In: 37th Annual Conference on IEEE Industrial Electronics Society (IECON 2011); 7–10 November 2011; Melbourne, VIC: IEEE; pp. 3198–3202. DOI:10. 1109/IECON. 2011. 6119822.

[18] Mishnaevsky L, Freere P, Sinha R, Acharya P, Shrestha R, Manandhar P. Small wind turbines with timber blades for developing countries: Materials choice, development,

installation and experiences. Renewable Energy. 2011; 36(8): 2128–2138. DOI:10. 1016/j. renene. 2011. 01. 034.

[19] Deublein D, Steinhauser A. Biogas from Waste and Renewable Resources. Weinheim: Wiley-VCH; 2008.

[20] National Academy of Sciences. Methane Generation from Human, Animal, and Agricultural Wastes. 1977.

[21] Office of the Leading Group for the Popularisation of Biogas (OLGPB) in Sichuan Province, Peoples' Republic of China. A Chinese Biogas Manual. 1978.

[22] U. S. Department of Energy. Microhydropower Systems [Internet]. 2012. Available at: www. energysavers. gov/your_home/electricity/index. cfm/mytopic=11050 [Accessed 2012-03].

[23] Alternative Energy News Network. Micro Hydro Power- Pros and Cons [Internet]. Available at: www. alternative-energy-news. info/micro-hydro-power-pros-and-cons [Accessed 2012-03].

[24] Oregon Department of Energy. Micro Hydroelectric Systems [Internet]. Available at: www. oregon. gov/ENERGY/RENEW/Hydro/Hydro_index. shtml [Accessed 2012-03].

[25] Wheldon A. Micro-hydro [Internet]. Available at: www. ashdenawards. org/micro-hydro [Accessed 2012-03].

[26] Maher P, Smith NPA, Williams AA. Assessment of pico hydro as an option for off-grid electrification in Kenya. Renewable Energy. 2003; 28(9): 1357–1369. DOI:10. 1016/S0960-1481(02)00216-1.

[27] Hravshat ES. Off-grid hybrid wind-diesel power plant for application in remote Jordanian settlements. Clean Technologies and Environmental Policy. 2009; 11(4): 425–436.

[28] Chen J, Che Y, Zhao L. Design and research of off-grid wind-solar hybrid power generation systems. In: 4th International Conference on Power Electronics Systems and Applications (PESA); 8–10 June 2011; Hong Kong: IEEE; pp. 1–5. DOI:10. 1109/PESA. 2011. 5982922.

[29] Vick BD, Neal BA. Analysis of off-grid hybrid wind turbine/solar PV water pumping systems. Solar Energy. 2012; 86(5): 1197–1207. DOI:10. 1016/j. solener. 2012. 01. 012.

[30] Brent AC, Rogers DE. Renewable rural electrification: Sustainability assessment of mini-hybrid off-grid technological systems in the African context. Renewable Energy. 2010; 35(1): 257–265. DOI:10. 1016/j. renene. 2009. 03. 028.

[31] Morea F, Viciguerra G, Cucchi D, Valencia C. Life cycle cost evaluation of off-grid PV-wind hybrid power systems. In: 29th International Telecommunications Energy

Conference, 2007 (INTELEC 2007); September 30 2007 to October 4 2007; Rome: IEEE; pp. 439–441. DOI:10. 1109/INTLEC. 2007. 4448814.

[32] Shaahid SM, El-Amin I, Rehman S, Al-Shehri A, Ahmad F, Bakashwain J, Al-Hadhrami LM. Techno-economic potential of retrofitting diesel power systems with hybrid wind-photovoltaic-diesel systems for off-grid electrification of remote villages of Saudi Arabia. International Journal of Green Energy. 2010; 7(6): 632–646.

[33] Ambia MN, Islam MK, Shoeb MA, Maruf MNI, Mohsin ASM. An analysis & design on micro generation of a domestic solar-wind hybrid energy system for rural & remote areas—Perspective Bangladesh. In: 2nd International Conference on Mechanical and Electronics Engineering (ICMEE), 2010; 1–3 August 2010; Kyoto: IEEE; pp. V2-107–V2-110. DOI:10. 1109/ICMEE. 2010. 5558476.

[34] Bekele G, Tadesse G. Feasibility study of small Hydro/PV/Wind hybrid system for off-grid rural electrification in Ethiopia. Applied Energy. 2012; 97: 5–15. DOI:10. 1016/j. apenergy. 2011. 11. 059.

[35] Khan MJ, Iqbal M. Analysis of a small wind-hydrogen stand-alone hybrid energy system. Applied Energy. 2009; 86(11): 2429–2442. DOI:10. 1016/j. apenergy. 2008. 10. 024.

[36] Kanase-Patil AB, Saini RP, Sharma MP. Integrated renewable energy systems for off grid rural electrification of remote area. Renewable Energy. 2010; 35(6): 1342–1349. DOI:10. 1016/j. renene. 2009. 10. 005.

PERMISSIONS

All chapters in this book were first published in GG, by InTech Open; hereby published with permission under the Creative Commons Attribution License or equivalent. Every chapter published in this book has been scrutinized by our experts. Their significance has been extensively debated. The topics covered herein carry significant findings which will fuel the growth of the discipline. They may even be implemented as practical applications or may be referred to as a beginning point for another development.

The contributors of this book come from diverse backgrounds, making this book a truly international effort. This book will bring forth new frontiers with its revolutionizing research information and detailed analysis of the nascent developments around the world.

We would like to thank all the contributing authors for lending their expertise to make the book truly unique. They have played a crucial role in the development of this book. Without their invaluable contributions this book wouldn't have been possible. They have made vital efforts to compile up to date information on the varied aspects of this subject to make this book a valuable addition to the collection of many professionals and students.

This book was conceptualized with the vision of imparting up-to-date information and advanced data in this field. To ensure the same, a matchless editorial board was set up. Every individual on the board went through rigorous rounds of assessment to prove their worth. After which they invested a large part of their time researching and compiling the most relevant data for our readers.

The editorial board has been involved in producing this book since its inception. They have spent rigorous hours researching and exploring the diverse topics which have resulted in the successful publishing of this book. They have passed on their knowledge of decades through this book. To expedite this challenging task, the publisher supported the team at every step. A small team of assistant editors was also appointed to further simplify the editing procedure and attain best results for the readers.

Apart from the editorial board, the designing team has also invested a significant amount of their time in understanding the subject and creating the most relevant covers. They scrutinized every image to scout for the most suitable representation of the subject and create an appropriate cover for the book.

The publishing team has been an ardent support to the editorial, designing and production team. Their endless efforts to recruit the best for this project, has resulted in the accomplishment of this book. They are a veteran in the field of academics and their pool of knowledge is as vast as their experience in printing. Their expertise and guidance has proved useful at every step. Their uncompromising quality standards have made this book an exceptional effort. Their encouragement from time to time has been an inspiration for everyone.

The publisher and the editorial board hope that this book will prove to be a valuable piece of knowledge for researchers, students, practitioners and scholars across the globe.

LIST OF CONTRIBUTORS

Jarotwan Koiwanit, Christine Chan and Paitoon Tontiwachwuthikul
Faculty of Engineering and Applied Science, University of Regina, Saskatchewan, Canada

Anastassia Manuilova and Malcolm Wilson
ArticCan Energy Services, Regina, Saskatchewan, Canada

Veerasamy Sejian
Icar-National Institute of Animal Nutrition and Physiology, Adugodi, Bangalore, Karnataka, India
School of Agriculture and Food Sciences (Animal Science) The University of Queensland, Gatton, QLD, Australia

Raghavendra Bhatta, Pradeep Kumar Malik and Bagath Madiajagan
Icar-National Institute of Animal Nutrition and Physiology, Adugodi, Bangalore, Karnataka, India

Yaqoub Ali Saif Al-Hosni, Megan Sullivan and John B. Gaughan
School of Agriculture and Food Sciences (Animal Science) The University of Queensland, Gatton, QLD, Australia

Maria do Carmo Rangel and Sirlene B. Lima
Grupo de Estudos em Cinética e Catálise, Instituto de Química, Universidade Federal da Bahia, Campus Universitário de Ondina, Salvador, Bahia, Brazil
Programa de Pós-Graduação em Engenharia Química, Rua Aristides Novis, Salvador, Bahia, Brazil

Sarah Maria Santana Borges and Ivoneide Santana Sobral
Grupo de Estudos em Cinética e Catálise, Instituto de Química, Universidade Federal da Bahia, Campus Universitário de Ondina, Salvador, Bahia, Brazil

Mohanned Mohamedali, Devjyoti Nath, Hussameldin Ibrahim and Amr Henni
Industrial/Process Systems Engineering, Faculty of Engineering and Applied Science, University of Regina, Regina, SK, Canada

Lilong Chai
National Engineering Research Centre for Vegetables, Beijing Academy of Agriculture and Forestry Sciences, Beijing, P.R. China
Department of Agricultural and Biosystems Engineering, Iowa State University, Ames, USA

Chengwei Ma
College of Water Resource and Civil Engineering, China Agricultural University, P.R. China, Beijing, P.R. China

Baoju Wang
National Engineering Research Centre for Vegetables, Beijing Academy of Agriculture and Forestry Sciences, Beijing, P.R. China
Key Laboratory of Urban Agriculture (North), Ministry of Agriculture, P.R. China, Beijing, P.R. China

Mingchi Liu and Zhanhui Wu
College of Water Resource and Civil Engineering, China Agricultural University, P.R. China, Beijing, P.R. China
Key Laboratory of Urban Agriculture (North), Ministry of Agriculture, P.R. China, Beijing, P.R. China

Bernardo Llamas
Escuela de Ingenieros de Minas y Energía, Universidad Politécnica de Madrid (UPM), Madrid, Spain
Inergyclean Technology, Almería, Spain

Benito Navarrete
Escuela Técnica Superior de Ingeniería de Sevilla, Sevilla, Spain

Fernando Vega
Inergyclean Technology, Almería, Spain

Elías Rodriguez
Iberdrola Generación, Madrid, Spain
Westec Environmental Solutions, llc, Chicago, USA

Luis F. Mazadiego and Ángel Cámara
Escuela de Ingenieros de Minas y Energía, Universidad Politécnica de Madrid (UPM), Madrid, Spain

Pedro Otero
Es.CO2. Centro de Desarrollo de Tecnologías de Captura de CO2, CIUDEN, León, France

Silvina M. Manrique and Judith Franco
Non Conventional Energy Resources Investigation Institute (INENCO) of National University of Salta (UNSa) and National Council of Scientific and Technical Research (CONICET), Salta, Argentina

Melanie L. Sattler
University of Texas at Arlington, Arlington, TX, USA

Index

A

Absorption, 46, 61, 81-82, 85, 120-122, 126-127

Adsorption, 57, 59, 62-67, 69, 72-75, 77-83, 88, 120, 122-123

Aermod, 1-2, 4, 11-15, 22

Air Dispersion, 2-5, 10-12, 15, 19

Air Dispersion Modeling, 2-4, 10, 15, 19

Atmosphere, 1, 11-12, 15, 29, 59, 115, 117, 122, 140-141, 150-151, 155

B

Biomass, 135, 140-141, 150-151, 153, 156-160, 165, 167, 169-170, 175-180, 182-186

C

Carbon Dioxide, 1-2, 18, 21, 24, 27, 41-45, 50-51, 53-57, 75, 84, 104, 117, 135-136, 138-142, 144-146, 149, 179

Carbon Dioxide Capture, 1-2, 18, 21, 54

Carbon Sequestration, 150, 153, 155, 158, 161, 165, 167, 169-171

Carbon Stock, 155, 158-167, 169-170

Ccs, 20, 23, 42, 117-118, 122, 148-149

Ccu, 42, 118

Cfpp, 99-100, 104-105, 107-112

Chemical Absorption, 120-121

Clean Biomass, 176-177, 180, 182

Climate Change, 2, 20, 24-27, 34-37, 39-40, 42, 54, 60, 99-100, 104, 114, 117, 148, 150-153, 171-173, 176-180, 185-187

Co2 Capture, 1-3, 10-11, 13, 15, 18-19, 23, 59, 61-67, 72-73, 75, 78-79, 81-85, 117-119, 121-122, 124, 126-127

Co2 Capture Technologies, 10, 15, 18-19, 117, 119, 124

Co2 Storage, 117, 127, 129-130, 133, 149

Co2 Uses, 117, 133-134, 136, 144

D

Deep Saline Aquifers, 129-130

Deforestation, 27, 150-154, 160, 170-171, 173, 177

Dehydrogenation, 41, 43-45, 50-51, 53, 55-58

Developing Countries, 26, 28, 37, 176-180, 182-184, 186-187

E

Ecosystem, 39, 151-155, 157-161, 163-168, 170-172, 174

Edge Effect, 150, 167, 169, 171

Emissions of Co2, 2, 60

Enhanced Oil Recovery, 42, 118, 135

Enteric Ch4 Emission, 24-25, 28-29, 34

Enteric Fermentation, 25-27, 29, 33, 35

Enteric Methane, 24, 27-30, 36-37

Esp, 1, 3-5, 11

Ethylbenzene, 41, 43-45, 50-53, 55-58

F

Fgd, 1, 3-5, 11

Fossil Energy, 100-101, 105, 108-110

Fossil Energy Systems, 100-101, 108-109

Fragmentation, 150-152, 155, 167, 170-171, 174

G

Gfpp, 99-100, 104-105, 107-112

Ghg, 2, 18, 24-27, 31, 33-35, 37, 99-101, 104-112, 118, 141, 148, 150-153, 157

Ghg Emission, 24, 33, 35, 100-101, 107, 109-110, 150

Greenhouse Gases, 1, 24, 36-38, 40, 55, 59, 99, 105, 175

Greenhouse Heating, 99-101, 103-105, 108, 112, 114

Gshps, 99-109, 111-112

H

Heavy Metals, 1-2, 5, 10-12, 14-20, 22

Human Health, 1-3, 5, 18-20, 22-23

I

Ionic Liquids, 61, 81-82, 127

Ipcc, 20, 25, 28-29, 35-36, 39, 54, 104-105, 114, 119, 152, 173, 178, 186

Iron Oxide, 41, 44, 46, 49, 55

L

Lci, 3, 9-10, 14

Livestock, 24-37, 40, 152, 160, 163, 167, 183

Livestock Farms, 25, 27, 33

Livestock Manure, 27, 31

Livestock-related Ghgs, 24, 35

M

Macrofouling, 146-147

Manure Management, 24-25, 31-33, 35, 39

Methane, 5, 24-25, 27-32, 36-40, 42-43, 55, 63, 75, 130, 133, 141-142, 144, 182, 188

Micro-hydro Systems, 183-184, 186

Microalgae, 135, 140-142

Microorganisms, 32, 141

Modeling, 2-4, 10-15, 19, 21-23, 25, 34, 40, 172

Mofs, 59, 61-62, 66-68, 72-75, 77-81, 83-86

N

Native Forest, 150, 152

Nitrous Oxide, 24-25, 32, 37, 39

O

Open Metal Sites, 73, 81

Oxy-fuel Combustion, 1-3, 10-13, 15-17, 19, 21, 123, 126, 148

P

Photosynthesis, 127, 135, 140-142

Post-combustion Capture, 11, 13, 17, 59-60, 63, 119-120

R

Renewable Energy, 42, 115, 144-145, 176-177, 179-180, 183, 185-189

Rumen Methanogenesis, 29-30

S

Sge, 99-100, 103, 107-108, 112

Shallow Geothermal Energy, 99-101, 110, 114

Solar Power, 23, 180

Solid Sorbent, 59, 62-63

Styrene, 41, 43-44, 51-52, 56-57

Sustainable Energy, 90, 114, 176-179, 185-187

Swhs, 180-181

U

Urban Agriculture, 99, 113

W

Water Stability, 77, 79-80, 84-86

Wind Power, 181-182, 187

Y

Yungas Ecosystem, 153-154, 160

Z

Zeolites, 63-65

Zifs, 59

Zirconia, 41, 44, 46-53, 58